CAVE EXPLORING

Other TAB books by the author:

No. 909 *How To Build Metal/Treasure Locators*
No. 1254 *How To Build Hidden, Limited-Space Antennas That Work*
No. 1316 *The Master Handbook of Telephones*
No. 1409 *Build a Personal Earth Station for Worldwide Satellite TV Reception*
No. 1419 *Astronomy and Telescopes: A Beginner's Handbook*
No. 1433 *Make Your Own Professional Home Video Recordings*
No. 1486 *32 Electronic Power Supply Projects*
No. 1487 *The Shortwave Listener's Antenna Handbook*
No. 1496 *The IBM Personal Computer*
No. 1506 *Making Money With Your Microcomputer*
No. 1526 *33 Games of Skill and Chance for the IBM PC®*
No. 1536 *Beginner's Guide to Reading Schematics*
No. 1556 *Graphics Programs for the IBM PC®*
No. 2097 *All About Electric & Hybrid Cars*
No. 2321 *The Joy of Flying*

CAVE EXPLORING

BY ROBERT J. TRAISTER

This book is dedicated to the memory of the
late Carlos Wine, former manager of Skyline
Caverns in Front Royal, Virginia. He was a vibrant
individual who was always willing the share his knowledge
with those of us fortunate enough to have known and worked with him.

FIRST EDITION

FIRST PRINTING

Printed in the United States of America

Library of Congress Cataloging in Publication Data

Traister, Robert J.
Cave exploring.

Includes index.
1. Caving. I. Title.
GV200.62.T7 1983 796.5'25 83-4885
ISBN 0-8306-0266-6
ISBN 0-8306-0166-X (pbk.)

Contents

Introduction

Since prehistoric times, man has entered underground caverns. Originally he was drawn to these alcoves by the protection caves offered from the elements. Cave entrances were also more easily defended. Later, man began to find that caves contained mineral resources and early mining became an art. During the Civil War in the United States, many southern caves were used as munitions caches and as living quarters for both Union and Confederate troops. This was a throwback to prehistoric times when the environmental protection and defensive possibilities of caves was utilized.

Although man has entered caves for many thousands of years, he did so with some trepidation. To the uninitiated, caves are mysterious and forboding. Caves are removed from many of our natural conditions, and our senses are not naturally equipped to handle such environments.

However, the human mind has allowed us to build and use implements that make the exploration and enjoyment of this vast natural resource practical and safe. Carbide lamps and other artificial sources of light take the place of the sun. People now enter caves not out of necessity, but with a desire to more fully understand what was previously unknown and surrounded by myth.

The modern cave explorer, often called a *spelunker*, usually thinks of caves and cave exploring in terms applicable to many of the more practiced sports such as football, baseball, and soccer. A cave exploration is a physical challenge but one that can be met and conquered with the proper planning and forethought. Spelunking is a pursuit that requires physical conditioning, mental alertness, and the practice of many different skills. Where else can you combine hiking, camping, mountain climbing, and other requirements of additional sports in one outing? The United States is blessed with a large number of caves and caverns, many of which are accessible to the individual who seeks a challenging sport that will not require great outlays of cash to enjoy.

On top of all of this challenge is the reward. The view of stalactites, stalagmites, and massive subterranean chambers filled with columns, flowstone, and other delicate calcite formations quickly transports one into a different world, one which seems to be completely disconnected from the earth as we know it. All physical laws still apply here, but they seem to have been interpreted differently. How else could something so far removed from what we know aboveground ever exist?

This book will take you through the basics of preparing to explore caves. No subject area has been left out. You will learn about the equipment involved, how to obtain it, and where it can be purchased. In-depth discussions will cover the use of this equipment. Underground scenes will be unfolded, and stories of subterranean explorations will be told. Most importantly, you will learn how to safely pursue this fascinating sport.

Cave exploring may not be for everyone, but it is a sport that can be successfully undertaken by literally millions of individuals who have not yet done so. It combines the best of two worlds—physical exercise and overpowering beauty. Once you have entered your first wild cave, you will want to do it again and again.

Chapter 1

How Caves Are Formed

Caves and caverns are formed over many hundreds of thousands and even millions of years primarily by water flowing across and into limestone. Caverns occur mainly in limestone because ground water circulates along the beds and joints and limestone is soluble in ground water. Typical limestone deposits may be anywhere from a few hundred to several thousand feet thick. The cave openings occur in the more soluble beds of this limestone; in some areas, large, weaker sections may be scooped away from the stronger surrounding wall in large chunks.

Most limestone is made chiefly of calcium carbonate, although other carbonates are also found. Limestone is dissolved when it comes in contact with water. Pure water will accomplish this, but when organic compounds and carbon dioxide are present in the water, the solution rate is many times greater. As discussed in the next chapter, carbon dioxide is absorbed from the atmosphere by rainwater and also from cave vegetation. As it travels through the soil, this forms a weak carbonic acid solution, which acts upon limestone to form the more soluble calcium carbonate.

Limestone is usually found in overlapping layers. As rainwater seeps through the soil, it moves downward along joints and beds in this material until it finally reaches the water table. These microscopic fissures are widened over thousands of years by the constant solution of the calcium carbonate. Many people think that a great body of rushing water forms caves through an erosion process created by the friction of the water against the limestone. This is a very great factor, but the solution of calcium carbonate is what causes the establishment of microscopic fissures which are gradually widened to a point where flowing water may enter and start the erosion process. The water flow enlarges the passageways through this friction effect, and most loose materials in the wall and ceiling are swept into the depths of the forming cave. As the flow of water increases, its solvent power along with its frictional wear effect becomes stronger. Small particles of stone and other debris in the water also increase the erosion of the limestone layer.

It is difficult to say just how long it takes to

form a cave in a particular area. This can be roughly estimated, but great accuracy cannot be achieved. Errors of several hundred thousand years are quite possible. The formation of a cave depends upon many, many different factors. First there is the thickness and makeup of the limestone. Limestone is composed of many layers of material, and naturally, some areas are weaker than others. Faults occur between layers and these can channel the flow of water at a much faster pace. Then, too, there is the amount of rainfall (over hundreds of thousands of years), the amount of carbon dioxide in the water, and the distance the limestone layer is beneath the earth. Most caves are formed in a similar manner, but each one is different and is formed over widely differing periods of time. Erosion by rushing water will depend upon the size of the channels that were formed by the slow solution of limestone. The quantity of water and the rate of flow, along with the resistance of the limestone, is often another big factor. Stream erosion can be identified in passageways which exhibit smooth walls and channels, undercut banks which create balconies and ledges, potholes along the walls and floors, and so forth.

Gravity plays a crucial role in the development of caves. Due to its effect, water always seeks the lowest level possible. The elevation of the limestone layer, then, plays a great part in the speed at which water flows. All ground water, as it seeps through the soil and into the limestone bed, seeks to reach equilibrium at sea level. For this reason, caves form much faster in mountainous areas high above sea level than they do at lower altitudes. In the Carlsbad Caverns in New Mexico, the limestone was deposited near the edge of an inland arm of the sea during Permian times, which date back to about 250 million years ago. Its core is a fossil barrier reef built by lime-secreting algae and other marine organisms. To the north are layered rocks which formed in a lagoon behind the reef. To the south are exposures of *talus*, or rock fragments, broken from the reef's crest by storms of the ancient sea.

In time, growth of the massive reef was halted and it became buried under layers of sediment. A pattern of cracks then appeared in the rock which set the stage for the formation of the caverns.

Rainwater, converted to a weak carbonic acid by absorption of carbon dioxide in the soil and decaying matter above, seeped into the cracks and worked its way down to the permanently saturated zone (water table). It then slowly dissolved the rock to create immense underground galleries.

As mountain building forces raised the caverns above the water table, air filled its chambers and mineral-laden water filtering in from the surface began to decorate the rooms with stalactite and stalagmite formations. A few are still slowly growing, but no change is noticed in a lifetime.

EROSION IN CHANNELS

The course of a stream may be influenced by greater solubility of the limestone, by a layer of impervious shale, or by fissures, which lead to lower channels. Two underground streams flowing in nearly parallel channels a relatively short distance apart at the same or approximately the same level may unite by the solution and erosion of the limestone wall between them. Examples of the convergence of two or more stream channels are found in many caverns.

Transported material and blocks of limestone from the cavern ceilings and walls are deposited in some rooms and along passages where tributary channels enter them. These deposits are generally covered with a coating of limy silt and flowstone, and broken travertine formations sometimes accumulate at such places. They may be cemented together by the growth of secondary stalagmites or by other travertine deposits. By the inwash of surface material through sinks, abetted by the collapse of weakened roofs, streams obstruct or partly destroy the mysterious underground channels they have created. Waters from relatively recent rains and floods may have clogged or partly dammed some cavern channels. When these partly clogged channels are again flooded, they are sometimes reopened.

Massive rocks dislodged by solution and erosion are found in Giant and Shenandoah Caverns in Virginia. In Shenandoah Caverns, huge slabs of limestone were loosened by water working along enlarged crevices believed to have been developed, in part at least, by fault movements. Large pendant columns and stalactites may be dislodged by the force of a large volume of water breaking suddenly into a channel. It is believed that large stalactite formations that have fallen from the roofs of some caverns are an example of this effect.

Barren places are found in many caverns. They are due mainly to the absence of crevices in the limestone, and hence the lack of ground water seepage and deposition. Some barren ceilings have resulted from the fall of travertine-covered slabs.

Many of the rooms and alcoves along the main channels at the junction of tributary channels are circular or oval, particularly where the tributary discharged into the main passage at such an angle that a whirlpool developed. At some places, circular channels have been carved in the ceilings, and fantastic designs have been etched in the ceilings and on the walls. Some of these features may have been formed in part by slow solution or by the gentle, ripple-like movements of standing bodies of water. Other peculiar pendant forms are thought to have been formed largely by stream erosion. Examples are found in most caverns, but are evident in Endless Caverns in Virginia. An interesting group of erosional features interspersed with stalactites, stalagmites, and flowstone can also be seen in Massanutten Caverns in Virginia.

BENCHES AND SHELVES

During the movement of waters charged with calcium bicarbonate through underground channels, thin films of calcium carbonate are deposited on the floors that, in time, may attain a thickness similar to that of the sheet and flowstone deposits described in the next column under Travertine Deposits. As the waters recede from parts of the old channels or abandon them for newer ones, these deposits dry out and crystallize into hard, protective coatings. Where streams have cut through these deposits and the fresh limestone beneath them has been dissolved, portions of the recrystallized layers are left as undercut benches or ledges. During repeated intervals of down-cutting and widening of the main channel, shelves or ledges have been formed at successively lower levels in most caverns.

TERRACES

By the deposition of a slimelike coating of flowstone over silt or clay deposits in rooms or channels, terraces are developed. Accumulations or broken travertine formations, fallen limestone blocks and inwashed material also are covered with similar coatings of flowstone resulting in terraces. In some caverns, a series of them has been built at successively higher levels. Terraces of variable extent and thickness occur in most caverns.

TRAVERTINE DEPOSITS

As long as the circulation of acidulated water continues, channels are enlarged and an extensive cavern is excavated. Uplift of the cavern area or sinking of the water table causes the channels to be abandoned. The further enlargement of these passages by solution and stream erosion ceases and deposition of travertine may commence in them. It is theorized that the change from solution excavation to depositional replenishment is brought about when the ground water which had completely filled the cavern during the progress of its excavation is drained away and its place is taken by ground air. The presence of the ground air provokes evaporation of percolating vadose water and escape of carbon dioxide from it, and thus compels it to deposit calcite and form dripstones on the cavern roof and floor. In a cavern system comprising channels and rooms at several levels, solution may be active in the lower channels while deposition is taking place in the upper levels.

Many attractive deposits of travertine can be found in caves. They vary greatly in size, shape, form, texture, and profusion. Some are translucent, many are tinted or streaked in bright colors, and some are resplendent with crystal brilliance. Crystals of calcite and veinlike stringers or bands of crystallized calcite are also encountered.

ORIGINS

The deposits of travertine, transforming plain and barren caves into veritable fairylands of bizarre and colorful forms, are composed of calcium carbonate that was dissolved from the limestone by percolating ground water. Common, drab limestone has thus been changed into travertine of great beauty. Deposition has been along the ceilings and walls as the water seeped through crevices and bedding planes in the limestone. It is caused mainly by evaporation of water in contact with air in the cave and by loss of solvent carbon dioxide from the water. The variation in shape, form and abundance of the deposits is due largely to differences in physical and chemical conditions obtained at the place and time the travertine was deposited.

Chapter 2

Cave Formations—From Stalactite to Stalagmite

One of the main beauties of cave exploring involves discovering the many unearthly formations that are created deep within our own earth. All cave formations are highly unusual by aboveground standards and often seem to hypnotize the neophyte spelunker. Even old-timers, however, never cease to be amazed at the beauty which unfolds upon entering nearly any cave. Some of the smallest caves have often produced the most beautiful formations, and each has its own calling card, as no two are exactly alike.

Cave formations are a result of the precipitate left behind by small drops of water dripping from the cave walls and ceilings a drop at a time for hundreds and thousands of years. No two formations are exactly alike, but all were formed in basically the same manner. You will find cave formations vary in size, shape, color, and form, and it's sometimes difficult to realize that a massive column or wall of flowstone was formed a microscopic bit at a time.

When it rains on the surface, the relatively pure water seeps into the ground and eventually exits into a cavern through billions of microscopic cracks in the walls and ceilings. During its trip from the skies into the caves, the water picks up a great deal of minerals. The water is mixed with carbon dioxide and this solution then becomes capable of converting calcium carbonate into soluble bicarbonate. The solubility property is determined by the amount of carbon dioxide in the ground water and also by the amount of organic acid present in this solution. As the carbon dioxide and organic acid increase, more calcium carbonate is absorbed.

The calcium carbonate is absorbed from limestone and the rate is much faster near the surface, decreasing as the water moves further underground. The two main mineral forms of calcium carbonate are calcite and aragonite. Both are identical chemically, but each crystallizes in a different manner.

STALACTITES

Stalactites are the overhead icicle-like projections which commonly exit from the cave ceiling. This term is often mispronounced because of another cave formation called a *stalagmite*. I have often heard stalactites referred to as "*stalagtites*." There is no such thing as a stalagtite; only a stalactite, or a stalagmite. Stalactites are formed by water filtering through the roof of the cave. Here it dissolves a small amount of lime carbonate in the ceiling. This then becomes solidified as further water is released, and a deposit of the lime carbonate is formed. In other words, the mineral content in the water only travels with the water. When the water evaporates, the mineral content is left behind. Remember now that we are talking about tiny microscopic deposits which have filled only a fraction of the total volume of a single drop of water. As this process continues over thousands of years, the precipitate ring builds up until a tiny nub is formed which is the beginning of the stalactite. If water continues to drop from the fissure, it flows down over this nub to its lowest point, where it adds more precipitate before falling to the cave floor below. The precipitate, however, is not added only to the tip, but to the sides of the stalactite as well. For this reason, the stalactite grows not only in length but in width as well. The pointed shape is still maintained, however, due to gravity's effect on the water.

Stalactites may be seen in great abundance in caves and often tend to accumulate in a few sections of the roof. These are the areas where most of the tiny cracks have occurred in the limestone, which allow water to enter the cave. Areas of the ceiling that contain few cracks will have fewer stalactites. The actual rate of growth of a stalactite depends directly upon the amount of water which is allowed to drip and the rate of the drip. A slow drip rate (although constant) will allow more precipitate to accumulate. A faster drip rate allows more of the carbonate to stay in the water, so the precipitate is not added as rapidly to the formation. Extremely slow drip rates do not build up much of a formation, nor do those which are exceedingly fast. The happy medium is somewhere inbetween, where the quantity of water is supplied at the proper rate to cause the largest addition of precipitate.

When you go cave exploring, you will undoubtedly see stalactites which are less than an inch wide at the base, the thickest part. You will also see some huge structures which may be several feet thick at the base and weigh a ton or more. Remember, both sizes were formed by the constant application of microscopic bits of calcium carbonate. Remember also that what you are witnessing will most likely be completely different from what a spelunker several thousand years from now will find in the same cave. The formation will be larger as drip continues, but could be gone if it stops.

To explain this further, remember that formations grow only when water continues to drip in the same location. When cave exploring, you will notice some formations that appear to be almost white and seem to glisten when struck by the glow of a carbide lamp. This glistening effect is a reflection of light from the water which is flowing on the formation. You won't be able to see the actual flow in most instances, as it is very slow, but occasionally, a drop of water will fall from the tip to the floor below. Other formations will appear to have very rough surfaces, may be gray in color, and do not exhibit the glistening effect. These are dead formations, in effect, or at least formations that are no longer growing. Due to some change (aboveground or below ground), water has ceased to drip from the fissure which originated the formation. No further deposits are being added and the formation is beginning to break down through the aging process. Since the formation was put there a bit at a time, it generally breaks down in the same manner, except in reverse. The formation deteriorates usually from the smallest point back toward the ceiling. You will undoubtedly see some stalactites which appear to have been broken off at the tip. This may actually be a result of the aging process. Rarely will a stalactite weaken at a point close to the base and then fall to the floor, although this *can* happen. I have never seen an example of a cave formation falling from the ceiling, nor viewed one that appeared to have already fallen in recent times. Stalactites are very sound structures. Small ones can be broken off by accidental contact with a spelunker's hard hat, but moderate-sized ones would probably require a sledgehammer. I have never been worried about being harmed by falling stalactites due to their mechanical makeup.

Fig. 2-1. Tiny stalactites, sometimes known as new stalactites.

Figure 2-1 shows a single tiny stalactite which is only a few inches in length and less than an inch in width at its widest point. This can be thought of as a new stalactite or baby stalactite, but its size may not be a true indicator of its age. This stalactite might have begun forming many thousands of years ago, but in the meantime it could have lain dormant for thousands of years when water ceased to drip through the ceiling in this area. At a later time, however, water could have begun dripping again, and its growth started anew. Generally speaking, however, smaller stalactites are much newer than larger ones. Remember, though, that the actual size of any cave formation is an indication of how long it has been growing based upon the rate of water flow and mineral content.

Figure 2-2 shows a massive stalactite which has been formed over many thousands of years and is still active in that water continues to flow over its surface. Such mammoth structures are not uncommon, even in some small to moderately sized caves, and are a daily sight of the active spelunker. It is difficult to imagine sometimes that such a large formation was developed by microscopic particles of minerals added in place one at a time.

STALAGMITES

Stalagmites are the mirror images of stalactites. Indeed, they most often form from water which has dripped from the end of a stalactite which lies directly overhead. Stalagmites are always solid cones of calcium carbonate which rise from the floor of a cave and sometimes from the horizontal portions of cave walls which form a collecting point for the overhead water drops.

When a stalactite is forming, not all of the minerals in the water which flows over it crystallize onto its surface. Many of them fall with the water

Fig. 2-2. A stalactite which is still active.

droplets to the surface below and crystallize here. As the water collects, the base is formed and begins to rise up from the cave floor. A small nub is formed, just as with a stalactite, and as water continues to drip from the stalactite above, it hits the top portion of the nub and then flows down over it, depositing precipitates evenly over the entire formation. The result of this action causes the base to widen along with the rest of the formation on a proportional basis. Due to gravity again, water flows down the side of the formation, so stalagmites are narrow at the top and wide at the bottom. In many instances, the stalagmite is a true mirror image and closely resembles the size, shape, and coloration of the stalactite overhead. This is not always the case, but is often so. Sometimes you will see a stalagmite without a matching stalactite overhead. This can be caused by a number of factors, including the breaking off of a stalactite due to a structural weakness. Often, however, the drip rate from the ceiling is too fast to form a stalactite, but the minerals collect on the cave floor where the water comes to a halt and the stalagmite forms slowly from this point. The space between the ceiling and the floor really has very little to do with stalactites and stalagmites

Fig. 2-3. Some moderately sized stalagmites.

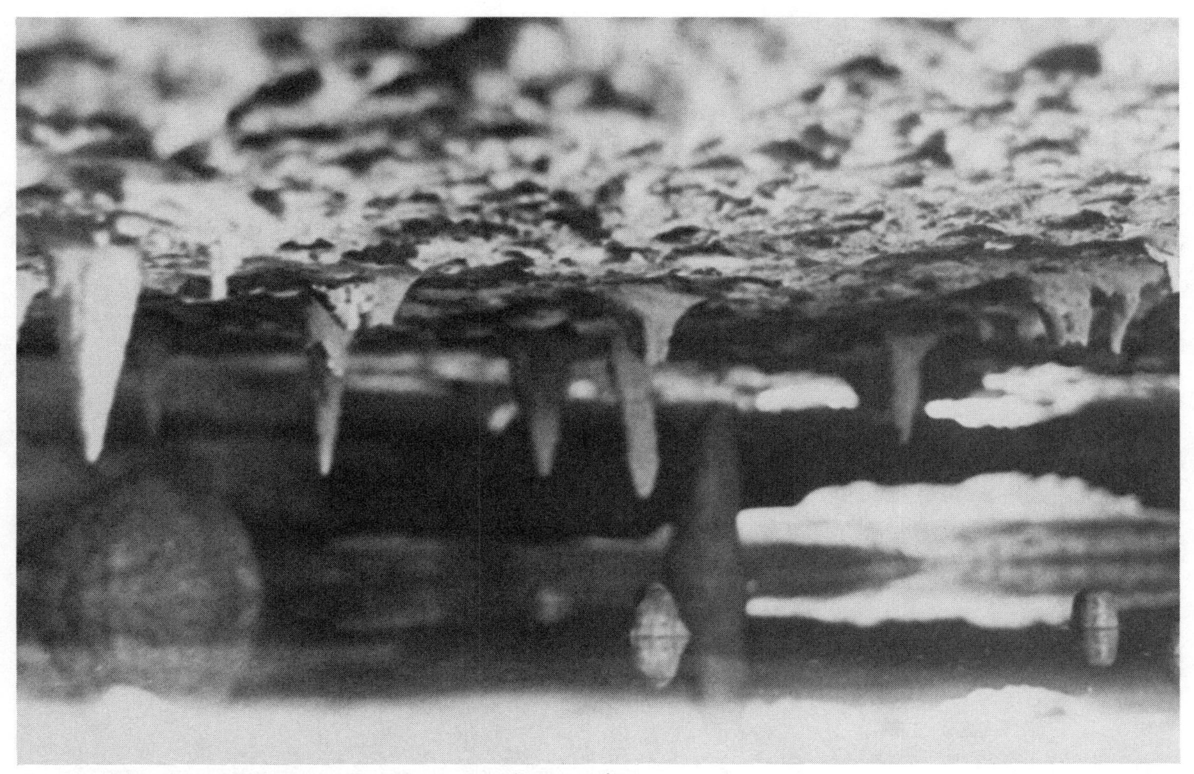

Fig. 2-4. Stalactite and stalagmite formations very close together.

forming. I have seen stalagmite formations on the floor of a large room whose ceiling was thirty or forty feet above. By close examination, the matching stalactite could be picked out far overhead.

Figure 2-3 shows some stalagmite formations of moderate size. Note their close resemblance to the stalactites pictured earlier. Figure 2-4 shows stalactite and stalagmite formations in a small area of a cave where the ceiling and floor are very close together. It can be seen that the overhead formation and stalagmite usually match up.

COLUMNS

How long can a stalactite or stalagmite continue to grow? The answer would seem to be as long as water continues to flow over them. This, however, is not the case in one sense, which is demonstrated in Fig. 2-5. This formation is called a *column* and results when a stalactite and stalagmite form for a long enough period of time to actually grow together. Obviously, the statacite continues to grow toward the floor, while the stalagmite grows toward the ceiling, and specifically, toward the stalactite overhead. Eventually, the precipitate buildup causes them to be separated by only an inch or so, and then the two fuse together, becoming one formation. Figure 2-6 shows a stalactite and stalagmite which are separated by only a short distance. Figure 2-7 shows what these two formations will probably look like thousands of years from now. This latter photograph shows a column which is fairly new by cave standards. Here, the stalactite and stalagmite have only recently joined together and still present a tapered appearance.

If water continues to flow across the stalacitite/stalagmite combination, the formation will continue to grow. However, the growth tends to be horizontal and the tapered effect is gradually filled

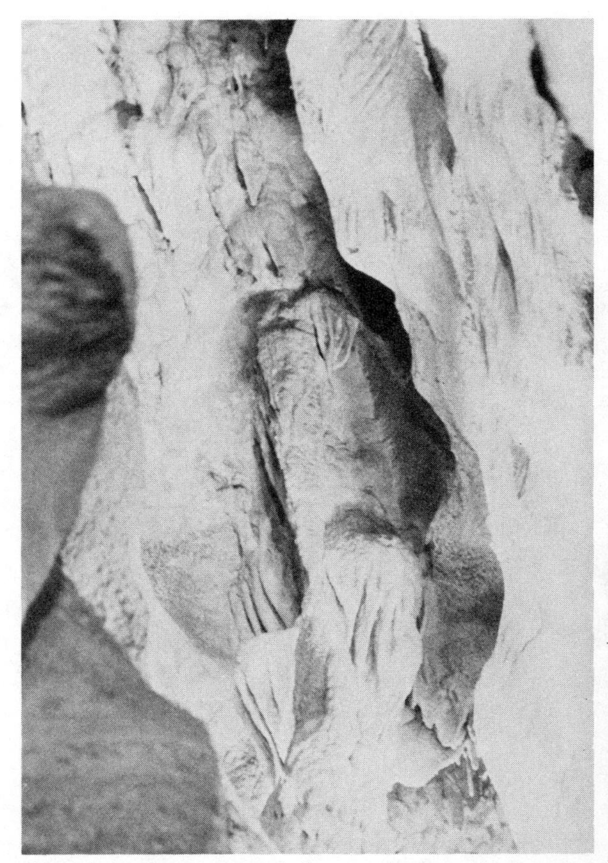

Fig. 2-5. A column is formed when a stalactite and stalagmite grow together.

in by the minerals, because they collect at the junction point. The bases of both formations become wider, but so do the other portions, and most columns are roughly uniform in width as they begin to age. Figure 2-8 shows a large column which was probably formed originally by the mating of two small stalactite/stalagmite formations. As water continued to flow, it became almost as fat as it was tall.

FLOWSTONE

Flowstone, again, is a calcium carbonate formation that is formed from the constant flow of water down the vertical surfaces of cave walls. Not all water enters a cave through cracks in the ceiling alone. Much of it seeps in through equally minute fissures in the walls. There is no actual dripping action here, so stalactites and stalagmites are not formed. The calcite is then left in a layer over the surface of the walls, precipitating in a rock surface which is smooth (if still wet) and has flow-like angles. Flowstone often forms on columns and other cave formations and is one of the more beautiful sights in most caves. Flowstone can also be found on cave floors where there is a slight grade to allow further flowing action. Figure 2-9 shows a large flowstone formation.

HELICTITES

A *helictite* is a highly unusual cave formation which seems to defy the laws of gravity. In describing the standard formations such as stalactites,

Fig. 2-6. Here, a stalactite and stalagmite are still separated.

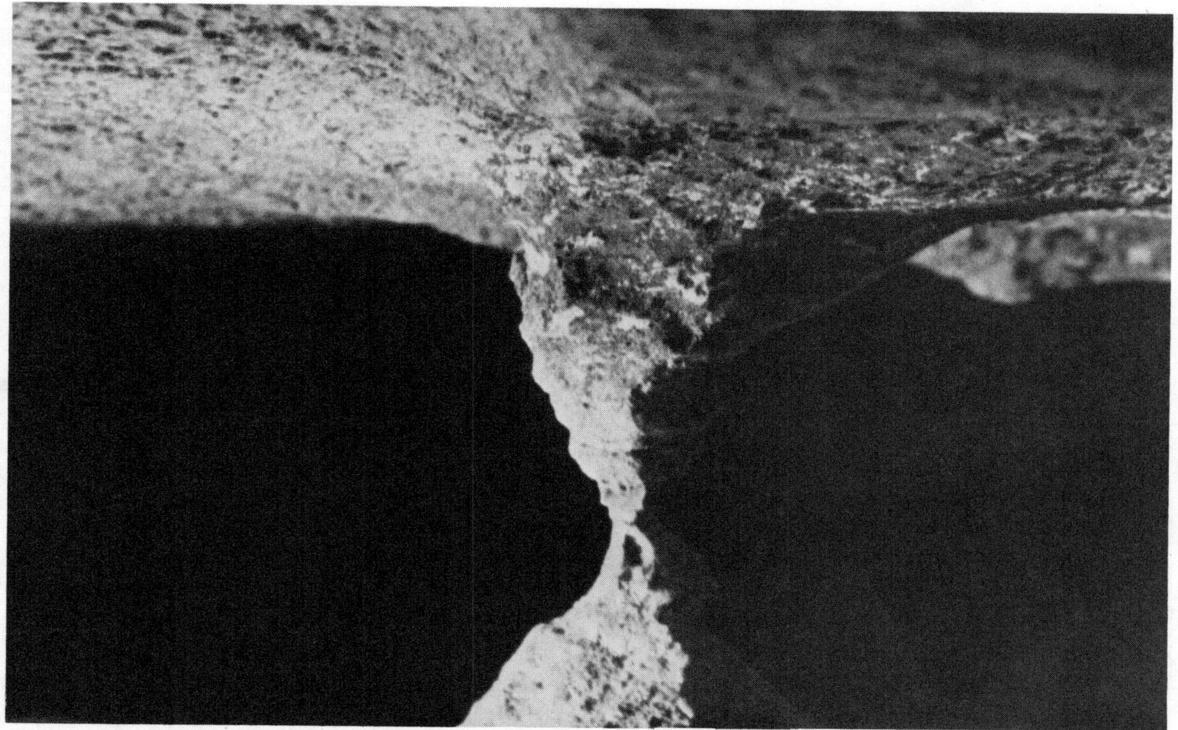

Fig. 2-7. Two formations have fused at this stage.

stalagmites, flowstone, and columns it can be seen how gravity is instrumental to their formation and appearance. Helictites, on the other hand, often twist upward and take on spider-like shapes.

Gravity does, however, play a role in their formation, and a helictate actually starts its growth as a stalactite on the cave ceiling. It may continue as a small stalactite formation for many thousands of years, hanging straight down from the ceiling. Suddenly, impurities appear in the water, changing the entire chemical composition. This causes the single formation to assume a conical shape. Each conical crystal fits into the other, much as you would fit several paper cups together. This is not a particularly sound mechanical structure, so as the crystals grow the whole formation shifts a bit, and the end moves away from true vertical. The formation has a tiny hollow tube running its entire length, so water will continue to flow due to back pressure until it emerges from the end. If the amount of water present at the larger base end of the tube is sufficient, then water will continue to be forced through it even if the opposite end is pointing toward the ceiling again. It can be seen that gravity still plays a crucial role even in the formation of helictites.

CAVE FLOWERS

Cave flowers are actually misnamed; they are not flowers at all, but crystallized mineral formations. While there are many different types of such formations, these are relatively rare and are especially precious finds for those spelunkers who happen upon them. I'm referring here to visual finds, as it is entirely inappropriate to remove them, or any other formation for that matter.

The rarest example of cave flowers is known as an *anthodite*. These have only been discovered at Skyline Caverns in Front Royal, Virginia. Shown in Fig. 2-10, anthodites are composed of thin calcite needles which jut out in all directions from the ceiling. These are still made from calcium carbonate, as are the stalactites and stalagmites, but they

Fig. 2-8. After many thousands of years, two separate formations have created this column.

Fig. 2-9. A large flowstone formation.

Fig. 2-10. Anthodites resemble tiny flowers. (courtesy Skyline Caverns)

Fig. 2-11. Another view of anthodites.

have crystallized into a different form called *aragonite*. The anthodites in Skyline Caverns are snow white but do not glow in the dark, as some imagine.

No one knows exactly how anthodites are formed, although some geologists feel that capillary action may be as good a theory as any. Capillary action deals with an absorptive effect, much like a sponge sucking up water. This might explain why these formations seem to defy the law of gravity. Water does not drip from these formations as it does on others found in caves, so no one knows for sure what caused them to appear in the first place, nor even if they're growing today.

The anthodites in Skyline Caverns were found in small air pockets located in a few of the many passageways in the commercial cavern system. Most of the passageways were ten to twelve feet in height but were filled with clay to within a few feet of the ceiling. For commercialization purposes, the clay has been removed from these rooms to allow access to visitors. Figure 2-11 shows another view of one of the anthodite rooms. Here you can see where the top layer of clay rested for many thousands of years. The anthodites around this are often colored with the iron oxide content of the clay. Those which lie an inch or so above this point are snow white.

Figure 2-12 shows an unusual anthodite formation which contains a small white stalactite at its center. This soda straw formation has grown out of the main anthodite and was obviously caused by dripping water which deposited the same type of crystalline structure (aragonite) that makes up the anthodites. This is quite possibly the most unusual formation in the entire world. Traveling further into the anthodite passageways a high ceiling is encountered, and cupped in one of the domelike openings is a striking collection of these beautiful formations composed of tens of thousands of individual needles. It is not known why anthodites have formed in

Fig. 2-12. This anthodite formation has a tiny stalactite at its center. (courtesy Skyline Caverns)

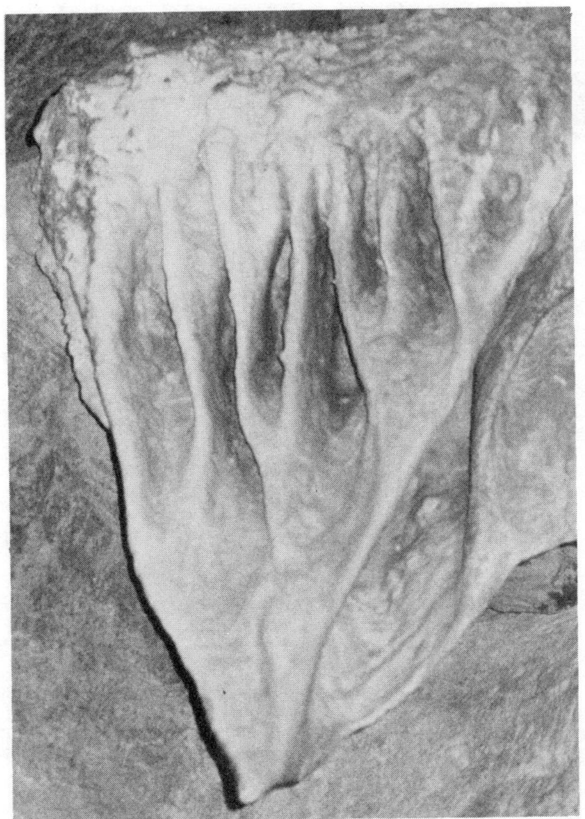

Fig. 2-13. A shield formation.

this cave alone. I have explored many other caves in the immediate vicinity of Skyline Caverns and have never seen another anthodite formation. Perhaps, however, if clay deposits are dug out of passageways in nearby caves, other anthodites might be found.

SHIELDS

A *shield* is a striking formation which is seen in some wild caves. Sometimes they are snow white, but most are the same color as the stalactites and stalagmites. A shield is a special type of flowstone formation which is made by water flowing through two flat disk formations which have become sandwiched together. Some shields are very large and may be in excess of twelve feet in diameter. Figure 2-13 shows an impressive shield formation in a commercial cavern system.

MISCELLANEOUS FORMATIONS

There are many other types of calcite formations which may or may not be encountered in the wild caves you explore. Calcite lily pads sometimes form in still underground pools from a localized buildup of precipitates on the surface. The water lines of some pools will even carry a thin, ice-like

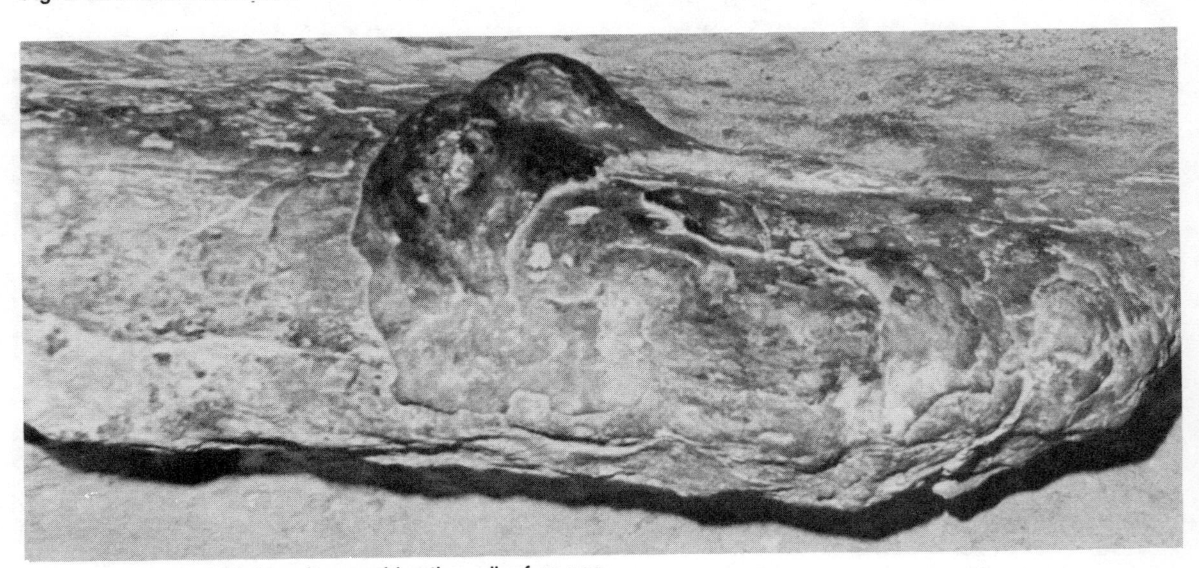

Fig. 2-14. This mineral deposit resembles the yolk of an egg.

layer of calcite which floats on the surface.

When stalactites or stalagmites are broken from the ceiling or floor of a cave, they sometimes leave a smooth mineral deposit which resembles the yolk of an egg. Figure 2-14 shows such an example. Other formations may look like popcorn, pearls, and even intricate maze configurations.

Most common formations in wild caves are stalactites, stalagmites, columns and flowstone. As a matter of fact, these will be seen in just about every cave you explore. White aragonite formations which resemble flowers are much rarer, as are shields, lily pads, and other specialized precipitate buildups. If you explore long enough, however, you may be able to witness all of these formations in their natural underground environment.

Chapter 3

Cave Life

What sorts of creatures live in caves? This is a difficult question to answer, because it is necessary to break down the various cave denizens into several categories: those that reside within the cave itself, those that live in the twilight zone area near the cave entrance, those that are permanent residents as well as those that are visiting temporarily. While any creature that is commonly seen in the entrances or depths of a cave is classified as cave fauna, the *true* cave dweller lives deep within the cave in the dark zone and may never see sunlight. True cave dwellers are usually completely different from any relatives which may be found above ground.

Cave creatures are broken down into three basic categories. *Troglobytes* are creatures which are permanent residents of the inner depths of caves and are seldom found in other biological environments. *Troglophiles* are composed of permanent or semi-permanent residents of the twilight zone area. These are capable of living in the dark zones of caves as well, but may also complete their life cycles in other biological environments. The last category is *Trogloxenes*, which are temporary inhabitants of the twilight zone areas of caves only.

We will deal first with troglobytes. The most obvious example of this class of creature is the bat. Bats are troglobytes because they are permanent dark zone inhabitants which are seldom found in other areas. True, bats may be seen aboveground during the early and late evening hours, but they don't usually reside here. Typically, they return to their subterranean homes shortly before daylight. Occasionally a few will be seen in the twilight zone area of some caves, but the greatest quantity will be found deep underground.

While many people feel that bats are the most numerous of troglobytes in most caves, this is not usually true. There are literally hundreds of thousands of tiny, microscopic insects which outnumber the bats a hundred to one. Other troglobytes include blind fish, blind crayfish, blind salamanders, and a host of other aquatic and semi-aquatic animal life. Troglobytes are further divided into two subcategories. The animals just mentioned (except the bat) may belong to either. An accidental troglobyte is an animal which, by accident, has ended up in the depths of a cave and is able to survive. Most of the aquatic animals named have been carried into caves by aboveground streams which eventually flow underground. Fish, crayfish, and salamanders have evolved on the earth's surface, but some were carried underground, possibly thousands of years ago, and became accidental troglobytes. As time passed, however, their offspring became permanent troglobytes, living their entire existence underground and never seeing the light of day.

SUBTERRANEAN CONDITIONS

At this point, it becomes necessary to think about the conditions within a cave or cavern. First of all, there is no light. Once you have traveled a few feet into the entrance and through the twilight zone, all light is shut out. Cave darkness is total; no optical effects whatsoever exist. You can't even see your hand in front of your face. Because all optical senses are useless, the need for eyes or other types of optical sensors no longer exists.

Another factor is the climate. Aboveground, some animals traditionally gear up for changing weather conditions by growing additional fur, hibernating, or even flying south for the winter. In a cave, however, all climatic conditions are fixed. The humidity stays at a high, constant figure; the temperature does not change in summer or winter, and air currents are either negligible or of a constant velocity. There are no seasons underground, just one continuous set of climatic conditions.

Another important point is the food supply. In most instances, there is far less food for most creatures underground than there is on the surface. The food supply can change with the seasons and with the amount of water which flows into the underground system. Generally speaking, however, food is at a premium for most types of underground residents.

Given these three conditions (no light, constant climate, and scarcity of food), true troglobytes evolve differently than their aboveground counterparts. First of all, there is no need for eyes, since there is no light to be seen. Although the process takes many thousands of years, the eyes atrophy through non-use, and through evolution, they are simply omitted many generations down the line. It is conceivable that the other senses the animal has and which can be of value in a setting of complete darkness become intensified by the same process.

Skin pigmentation is a built-in safety feature for aboveground animals (including man) and serves to shield delicate skin tissue from the direct rays of the sun. Below ground, however, this is not a factor, so most true troglobytes are albino. Nature tends to get rid of anything which goes continuously unused.

Due to the relative food shortage, most permanent troglobytes are very small when compared to similar creatures found aboveground. Cave fish, for example, may be only an inch or so in length when fully grown, whereas aboveground they may be ten times that size. A large creature forced to live all of its life in an underground cavern would quickly starve to death. Small creatures have the ability to feed on tiny insects and microscopic organisms which are in more numerous supply, thus assuring their survival. Evolution takes care of the size factor, and it can be assumed that the original aboveground creature which was somehow washed into the underground was in the early stages of development when this occurred. For instance, a normal fish which suddenly found itself underground would not mature to the same size as it would aboveground due to the minimal food supply. Its growth would be stunted, and if it reached maturity, its offspring would reflect this restricted growth process as well. After a thousand years or so, a miniature edition of this original fish is the normal outcome.

The third factor, controlled climate, is theoretically a bar to a continuous evolutionary process. On the surface of the earth, such factors as ice ages, hot spells, climatic changes, and so on all contribute to normal evolutionary development. Underground, however, these factors are eliminated, so it could be theorized that once a creature becomes ideally adapted to a cave environment, it ceases to develop further. Severe climatic conditions, however, can affect some subterranean caverns by filling them with water, clogging all air passages with mud and debris, etc. These changes tend to kill off all subterranean life. When and if the cave returns to normal, the entire process would have to begin again from scratch, starting with some surviving aboveground animals being washed into subterranean chambers. Science fiction movies have played up the idea of creatures being washed into remote underground abysses and remaining the same for millions of years. They surface again in modern times and wreak havoc upon the earth—Tokyo has been completely devoured at least 300 times by Godzilla and similar counterparts. It must be remembered that while caves are highly isolated from the surface of the earth, they are still a part of this planet and can be affected by global changes. These effects often take a long time to occur, lagging behind their aboveground influence. However, once an upheaval is over and conditions return to normal on the surface, this return to normalcy may take far longer underground. In both instances, there is a lag of effects in the underground.

BATS

Every cave has the potential to harbor bats. Some small caves may hold only one or two, while others may be the home of thousands or even millions. Caves and bats are usually considered together in the minds of most people, and neither are very attractive. Since medieval times, bats have been portrayed as deadly carnivorous creatures which stalk the night skies, seeking out hapless victims. The bat has been feared probably for the same reason that many people fear caves . . . the unknown.

The bat is quite an unusual animal and might even be considered a freak in the mammal sector of the animal kingdom along with the duck-billed platypus. The bat is the only mammal with the ability to fly. Sure, there are flying squirrels (misnamed) and a few other rodents that are equipped with thick flaps of skin between the front and back legs which allow them to form a rather efficient airfoil in order to *glide* from a lofty perch to the ground below. However, these animals do not have the ability to fly as bats do. They cannot become airborne from the ground and cannot sustain themselves for more than a few seconds in the air.

On the other hand, the bat can become almost immediately airborne while resting on the ground. This creature is an excellent flier and is capable of many aerobatic maneuvers that even some birds can't attempt. Bats are also ancient animals. Scientists have checked fossilized remains and found that the bat dates back at least sixty million years.

There are two basic families of bats: those who eat insects and those who eat fruit. Here in North America, the bats we see are insect eaters and belong to the suborder *Microchiroptera*. The fruit eaters or *Megachiroptera* are mostly found in South America and are larger than the insect eaters of North America. There is one member of the bat family which confuses this order a bit, however. The vampire bat belongs to the insect eater category but doesn't eat insects. It doesn't eat fruit either, but subsists solely on the blood of animals. I'll bet this last statement doesn't do much to attract the newcomer to cave exploring, who is teetering on the edge of taking up this sport. Most people think of vampire bats as tremendously large and vicious creatures of the night. Nothing could be further from the truth. Vampire bats are larger than many insect eaters but certainly smaller than every fruit-eating bat. They don't swoop out of the sky and go for the throat with vicious six-inch fangs. They have tiny hollow teeth which are extremely sharp, and their saliva forms an anticoagulant which tends to keep the puncture wound open and the blood flowing. Vampire bats mostly take blood from cattle and other medium-to-large sized animals. They have been known to enter a campsite and even feed on the blood of human beings. You may be surprised to learn that the majority of people who have been bitten and actually fed upon by these bats were never aware of it until they awakened the next morning and discovered the tiny puncture marks. Most of the time, the bats go for the unprotected feet rather than the throat, and in any case, the bite is very minor.

Vampire bats are extremely dangerous creatures, however, because they are known carriers of rabies. While their bite does little direct physical harm, the indirect effects can be fatal.

Now that I've told you the horrible tale of the vampire bat, remember that this creature is native to South America; it is almost never seen in the United States and then only in the extreme southwestern part of the country near the Mexican border. The question will still arise occasionally about the possibility of encountering some mammoth-sized bat in a subterranean chamber. Barring a nuclear isotope mutation or some other horror from science fiction movies (the type seen only on third-rate late-late shows), there is no chance of such an encounter. There are some bats which are huge and have wing spans of five feet or more, but they subsist on a diet of fruits and plants and are not found in caves, at least not in North America. Even these large animals are timid creatures and avoid man whenever possible.

Most North American bats are tiny creatures. The first one I saw in a small cave in Warren County was not immediately identified as a bat. It was hibernating on the cave ceiling during the later winter months and looked more like a furry little worm. Its body was no longer than an inch and a half and would have gone quite unnoticed in most instances. This is a species which is commonly known as the little brown bat; it is quite common to North America, especially in the eastern part of the country, and subsists solely on insects. Most of these insect eaters must devour large quantities of insects every day. The total daily consumption of a well-fed bat will amount to more than half its full body weight. For this reason, bats are quite important to the ecological system and rid the world of large quantities of harmful insects.

Most types of bats are social creatures and live in large communes deep underground. They can often be seen in caves during the daylight hours in large groups on the ceiling, where they sleep until nightfall. A built-in alarm clock wakens them at dusk, at which time they exit to the surface above for a night of frenzied insect feeding. Some caves which contain a large number of bats make interesting observation at dusk, when bats can be seen exiting in great numbers, sometimes by the hundreds of thousands.

Nearly everyone knows that bats navigate by some form of physiological radar. This is not quite correct, but gets the point across. Bats actually navigate by a system which could better be classified as sonar. They are cable of emitting from their vocal cords a fast series of ultrasonic beeps. When these sound waves strike another object, they are reflected back to the bats' ears where detection occurs. The tiny bat brain is capable of extrapolating all kinds of information from the returning beeps: if something is ahead, how large it is, and possibly even what it's made of. If you have ever seen thousands of bats entering or exiting a cave, then you must know that their navigation system is quite efficient, because each one seems to be flying its own erratic pattern. The bat depends upon this system to avoid contact with other bats and, of course, stationary objects which lie in its path. Its highly advanced flying ability allows it to spin, loop, and weave in a split-second in order to avoid collisions during its relatively high-speed flight.

The saying "blind as a bat" is a complete falsehood. Bats have eyes and indeed, they use them when feeding on the surface. Their eyes have never been phased out by evolution, because these creatures do not spend 100 percent of their lifetimes in the absolute darkness of underground caverns. They have a definite need for eyesight and have retained it, along with the ability to navigate and exist in complete darkness for many millions of years. "Blind as a bat" probably originated because these creatures have often been known to fly directly at an object or a person, veering away only at the last minute, leading some people to believe that bats cannot see where they are going until they are a few inches away from the object. Here it would seem that "nearsighted as a bat" might be more appropriate, assuming that the supposition was true in the first place. Vision has nothing to do with it, however. Their built-in sonar allows them to sense a distant object, judge the distance, and then take the proper corrective maneuver. The fact that this is done in close proximity to the object doesn't make any difference to the bat, so it should not affect you as a spelunker either. Chances are you will eventually find yourself in a position where bats are flying down the same passageway you're traveling. This is nothing to be really concerned about, although if there are large numbers of them you might want to protect your eyes by simply holding your arms in front of your face, particularly if the passageway is small. The bat's navigational system is very good, but it is not perfect. In cramped quarters and where the animals may be a bit panicky due to human presence, a minor collision between bat and spelunker will occasionally occur. I have been struck once or twice by a bat wing but have never experienced a full head-on collision. The bats certainly weren't attacking; they were just trying to get past me to the cave entrance (exit), and the largest opening available was not quite big enough for both of us. In any event, the presence of a large number of bats fluttering all around you is quite bothersome, and for this reason sleeping bats should not be disturbed. You can examine them from a distance as they hang upside down on the cave ceiling, and one or two may even open an eyelid and examine you back. If you keep a respectable distance, however, most bats will remain in place and a potential problem is avoided.

To repeat, the bat's navigational system is not perfect. When a bat enters the depths of a deep cavern through a maze of passageways, it is theorized that a bit of memory on the part of the animal is required for it to get back out again. Occasionally, however, bats will become hopelessly lost and will eventually starve to death. Due to their high-speed flying ability, they can check out many different passageways in a short period of time, and this hit-or-miss method is almost always adequate for them to find the proper exit channel. But like any system, failure can occur. The moral to this story is: If a *bat* can get lost in a cave after millions of years of experience underground, a spelunker is certainly capable of doing the same. (Chapter 6 deals with this subject, although we human beings are much smarter than bats and have few problems in this area.)

The construction of the bat's body is unique. The wings are thin membranes which are supported by tiny bony structures which at one time were the creature's fingers. These extend from the wing root to the very tip. They have the ability to move their wings at a wide variety of angles. This accounts for their maneuvering ability. Some of the latest aircraft now have adjustable wings which allow them to be converted from supersonic flight configurations to a wing design best suited for slow-speed maneuvering. The bat has had this ability for sixty million years.

As magnificent as the bat's wings are, the rest of his body is pretty typical as far as rodents are concerned. The main portion of its body is not especially designed with aerodynamics in mind, and the hind legs are useless when the creature is in flight. These rear appendages, however, are unique in that they allow the bat to cling to the cave ceiling. The grip of the claws is very strong and must be connected with a type of reflex action. Naturally, the grip must continue while the bats are in deep hibernation during the winter months. Bats which have been long dead may even be found still clinging to a cave ceiling. Once the grip is locked, it becomes automatic and can only be unlocked through a conscious effort on the part of the animal.

Many North American bats hibernate in caves during the cold winter months. A few northern species will, however, fly south for the winter, enjoying active feeding the whole year round. Hibernating bats depend upon a layer of fat which is stored during the feeding period of the warmer months. During hibernation, the bat will lose more than one-third of its total body weight. Scientists have found that bats will begin to emerge from caves around the start of spring, but in some colder climates they will emerge only on the warmer nights and then return to their underground chambers for a few more days or weeks of hibernation. This is quite unusual; when most mammals are awakened from hibernation, they cannot return to it. Many mammals which are accidentally brought out of their deep sleep before winter is over will die as their metabolic processes step up, burning away the stored fat. This is not true of bats, however. If they are disturbed during the winter months, they will simply go back to sleep again.

Bats mate before going into hibernation, but due to the almost nil bodily processes during this sleep period, the gestation time is delayed. It begins again when the bats leave the hibernating state, and in most instances, the young bats are born during the month of June. When the young bats are born, they are about 30 percent of the size of the adults, and their claws are completely developed, allowing them to cling tightly to their mother's fur. The young bat travels with his mother at all times, holding on for dear life, even while she is flying in search of food. Bats will typically give birth to two or three offspring during each delivery, although some have been known to have four or more. The young are not able to fly at birth, but approximately one month later they seem to be fully versed in even the most complicated maneuvers. It was mentioned earlier that bats are capable of moving their wings in many varied positions. Their normal flying stance, however, is completely different from that of birds. The movement of a bird's wings during flight is mostly confined to the vertical. However, a bat often moves through the air in a swimming motion using its elongated fingers and membrane to scoop into the air ahead and push it behind. This is a most unusual flying configuration and is one of the reasons for the bat's extreme versatility in the air.

Gregarious bats tend to live in colonies and hibernate together during the winter. They usually gather in clusters in caves, hollow trees, or other suitable roosts. There is a good reason for this; as temperatures drop and winter comes, the bat's body heat also decreases until it is just slightly warmer than the air. Clusters of bats, their bodies packed tightly together, create a mini-climate that ensures their survival. They store up an amount of fat sufficient to sustain them over the winter. If disturbed while hibernating, bats quickly rouse and use up stored nutrients. Since there are no insects to catch in the winter to replenish their fat, they may die before spring.

Solitary bats tend to roost in trees hanging from branches head-downward or hiding under loose bark. They are migratory, moving south late in the fall and north in the spring.

Bats have very sharp teeth to eat small insects but only the largest species (i.e., the big brown, hoary, red, and silver-haired bats) can bite hard enough to break human skin. Bat bites should be washed immediately for at least fifteen minutes, preferably with soap and hot water; a physician should also be called. The important thing to remember is that a bat flapping around on the ground may not be a healthy bat. People, especially children, should not go near it. However, should a person or a pet animal be bitten by the bat, it should be collected by gloved hand, stick or forceps and local public health authorities notified and informed of all circumstances.

Rabies is carried by bats, but not every bat may be rabid. Raccoons, skunks, squirrels and dogs also carry rabies. People who have been bitten by bats probably have disturbed or provoked them.

Scientists have noticed a serious decline in bat populations. Three species are on the Official U.S. Endangered Species List: the Indiana bat, the gray bat, and the Hawaiian hoary bat. Reasons for their decline include: Loss of cave habitants by an increase in surface mining operations and urbanization, cave commercialization, construction of dams resulting in flooding of caves, burning of debris in cave entrances and disappearing mine shafts and old hollow trees.

The second cause for bats' decline is vandalism. Many people, not realizing the importance of bats in the eco-system, kill them on sight. People, exploring caves and mines, disturb them accidentally or kill them for "fun." Disturbance of bats in roosts, wherever they may be, is another cause of their decline.

OTHER CAVE LIFE

There are many other types of small animals that may be found in caves, especially in the twilight zone area. Many of these are not true cave fauna and consist of flies, gnats, and crickets which are normally found only on the surface, but which have retreated to an existence (often temporary) in the cool recesses of cave openings. Most of these, however, will not normally be found far underground unless they have arrived there accidentally. Worms, mites, and aquatic insects are often swept far underground by waters rushing in from the outside. Some of these are able to survive continuously below ground, but many others will die. After a heavy rain, one often finds an abundance of small marine animals that normally live aboveground in cave streams. Most will die within a short period of time if not carried out again by the same stream.

The microscopic animal kingdom underground is quite abundant and varied. Tiny snails and beetles can sometimes be discovered under the lens of a microscope. Many years ago, Skyline Caverns in Front Royal, Virginia, discovered a microscopic snail bearing the scientific name *Lartetia Claustra Morrison*. At the time of discovery, the only other place in the world these snails were known to exist was in caves in the Rhine Valley in Germany. A microscopic beetle was also discovered, again with a very long scientific name, *Pseudoanathalamus Petrunkevich Valentine*. (In both cases, the last names of the creatures were derived from the name of the discovering scientist.) At the time of discovery, this eyeless beetle was known to exist nowhere else in the world.

PREHISTORIC REMAINS

Prehistoric remains are sometimes unearthed in many large caverns. Most often these are thousands of years old and are not the skeletons of true cave dwellers. Rather, these are prehistoric deer, felines, etc., which entered or fell into the cave and subsequently died there. Remains of prehistoric man have also been found in a few rare instances, especially where a particular cave was used as an ancient tomb.

Generally speaking, all true and permanent cave fauna originated aboveground and were swept underground by an accident of nature. Here, the creatures evolved according to a different set of guidelines, all determined by the fixed environment of subterranean passageways. The average spelunker will see bats, camel crickets, and little else in the deep interiors of caves. Occasionally a blind fish, crayfish, salamander, etc., will also be seen, but these occurrences are usually rare. This is partially due to the scarcity of this type of life in most caves and also to the fact that any types are very wary and can detect an alien presence long before the eyes of the spelunker detect them.

Chapter 4

The Tools of the Trade

As is the case with most pursuits, the equipment used for cave exploring activities will directly relate to the safety and success of each and every expedition. While the success of an outing and the fun of just being underground are important, the safety aspects cannot be overemphasized. For this reason, it is necessary to delve deeply into spelunking equipment.

Right from the start, it should be pointed out that there is no exclusive supplier of caving equipment and accessories that I know of. As a retail business, supplying the needs of cave explorers alone would not be a very lucrative profession. Fortunately, however, most of the equipment you will need as an amateur speleologist can be supplied by your local hardware and sporting goods outlets. Some of the equipment you can build yourself. However, most will be common items that may need to be modified to some extent for greater suitability for spelunking.

In this chapter we will not go into great detail on special climbing equipment needed by caving professionals, who may find it necessary to descend, by Alpine techniques, sheer cave walls hundreds of feet high. This entails specialized mountain climbing equipment, a discussion on which could easily fill several large volumes. The purpose of this book is to acquaint you with the fun of cave exploring and to provide you with the information to get started and advance to more difficult caves. For this reason, the basic necessities will be discussed—how to build or obtain them and how they should be used in underground exploration. Additionally, some accessories are also presented which will be used in specialized explorations.

HARD HATS

The first item of necessity for the cave explorer is the hard hat. This is an essential item that must *never* be neglected. Figure 4-1 shows the hat most popular for cave exploring purposes. This is a standard construction hard hat and contains a small frontal brim. Hats may be constructed from many different types of hard plastic or fiberglass and are commonly seen in the construction industry. If you have to purchase one new it may cost $30 or more,

Fig. 4-1. A standard construction hard hat is the most common type of headgear used for cave exploring.

but by checking at local construction sites, you should be able to pick up a number of used ones for less than $10. Retired construction workers often have a large collection of hats, more often than not relegated to a dusty attic or damp basement. When I began my cave exploring pursuits, I was fortunate to have a parent who had been in the electrical construction business for over 30 years, and these hats were a common fixture at home.

Figure 4-2 shows another type of common construction hard hat. This one contains a narrow brim around the entire circumference of the helmet. Although this type can be successfully used for cave exploring, it is bulkier than the other type, and the 360° brim is more apt to become wedged between narrowly spaced walls. Both types are usually fitted with an adjustable cloth or plastic webbing, which fits snugly over the head and also separates the hardened portion of the hat from the skull by an inch or so. This provides additional protection should the top of the hat be struck by a falling object.

If you can't locate a conventional hard hat, a football helmet will do. You can also use a motorcycle helmet or even a surplus military helmet. These are not preferred, however, as they are often heavier than hard hats, and some fit down over the ears and obstruct normal hearing.

Regardless of the type of hard hat you use, make certain that it is undamaged. Even a tiny crack or fissure can weaken the entire structure and cause it to give way to the sudden mechanical shock which can occur from a falling object. Hard hats have undoubtedly saved many lives and prevented untold thousands of concussions, skull fractures, and other cranial injuries. In many ways, your hard hat is your lifeline to cave safety and must be worn before entering the cave.

Never outfit any type of hard hat worn for cave exploring purposes with a chin strap. Should your hat become wedged in a tight crevice, the chin strap will prevent you from pulling your head out easily so that it can be removed. Also, should you happen to fall, your hat could become wedged and the chin strap could even strangle you. It is also for this reason that football and motorcycle helmets are not to be preferred even when the chin strap is eliminated. These fit very snugly on the head and overlap the ears. They are extremely difficult to remove with one hand, and you'll need both at all times when cave exploring. It is possible to use a hacksaw on the ear flap portion of a football or motorcycle helmet to remove these unwanted side pieces, but the fit then becomes very loose and jagged edges are often left, further adding to their unsuitability.

Construction hard hats are very rugged and will stand up under a lot of abuse. This is what they were made for. However, it is essential that after each expedition they be checked for any sign of damage. In some ways a damaged hat is worse than none at all, because the wearer depends on it to perform like an undamaged one and it can fail at the worst time. The hat I used in exploring my first cave over 20 years ago is the one I still wear today, and it would seem to be in almost perfect condition except for a large number of scratches that have accumulated over the years. A hard hat requires little maintenance, although it will be necessary to replace the internal webbing from time to time. It's always a good idea to pack an extra hat or two when heading out on any expedition. These are usually too bulky to be conveniently carried into the cave, but they can be stored in the trunk of an automobile ready for action should some member forget his hat or if a damaged hat is discovered.

CARBIDE LAMPS

We will return to the hat a little later on, but for now it is necessary to talk about the carbide lamp which mates with the hat during explorations. The carbide lamp is the universal light source for cave explorers as well as for miners and others who work in underground environments. It contains no batteries or incandescent bulbs that can chatter due to moderate mechanical shock. It needs very little attention while in operation, and is so simple that what little maintenance is needed can be performed in a few minutes. Figure 4-3 shows a typical carbide lamp which was purchased at a local sporting goods outlet.These are sometimes worn by fishermen and other sporting enthusiasts who need both hands free while plying their trades at night. The lamp is actually a miniature acetylene torch which operates from calcium carbide and water.

Calcium carbide looks very much like ordinary driveway gravel. Its odor is rather pronounced—an almost smokey smell. While unassuming in appear-

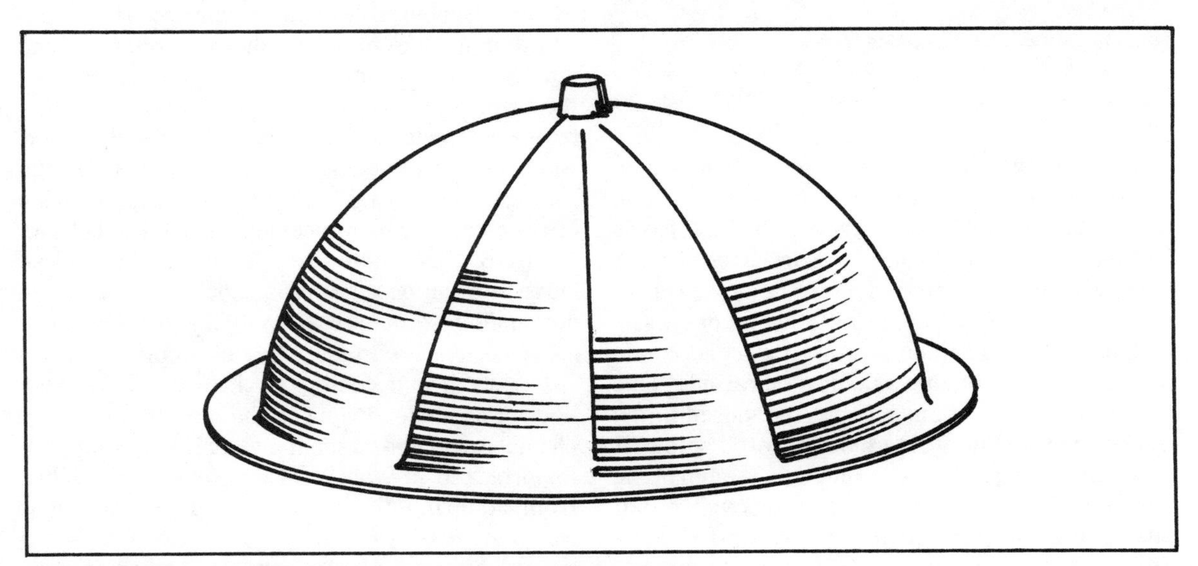

Fig. 4-2. This type of hard hat has a narrow brim around the entire hat.

Fig. 4-3. Carbide lamps are most often used to provide illumination.

ance, when a drop of water comes in contact with a single piece of calcium carbide, a chemical reaction occurs and acetylene gas is given off. This gas is ignited as it exits a miniature nozzle and produces a flame which is concentrated by an aluminum reflector. Figure 4-4 is an exploded drawing of a basic carbide lamp. It contains two main chambers, one for water and the other containing the carbide. This is a gravity-fed system, so water is housed in the top chamber, while the carbide is stored below. A small, adjustable valve is mounted on the top of the lamp (just behind the reflector); when triggered, it allows a small quantity of water to drip steadily down on the carbide in the chamber below. Acetylene gas is immediately given off and travels upward through a small metal tube to the center of the reflector. Here it exits through the tiny opening in the nozzle, where it is ignited to provide underground light. Most of the time, a striker mechanism is mounted inside and to one edge of the reflector. This is the same type of striking mechanism present in most cigarette lighters. A tiny piece of flint comes in contact with the iron striker wheel, creating a spark which ignites the accumulated acetylene gas.

Acetylene gas? Isn't this what is used in commercial torches which cut through one-inch steel plating? Yes, but the carbide lamp is a low-pressure system while commercial torches mix pure oxygen with the acetylene gas to produce a high-temperature flame. The flame produced by the carbide lamp is designed for illumination purposes only and can do little more than burn paper or an improperly positioned hand. With the small amount of gas produced, you need not fear any kind of explosion.

When dealing with any type of flammable gas, however, certain safety precautions are mandatory. The main precaution when using a carbide lamp is to make certain that the rubber gasket which attaches to the threaded lip of the carbide chamber is in good shape and properly secured. To load carbide in the bottom chamber, the whole assembly is simply unscrewed just like removing the cap from a canning jar. The rubber grommet forms a gas seal between the two mating surfaces, allowing all of the gas to travel up the tube and through the nozzle. However, when the gasket begins to deteriorate with age and use, gas can escape and will often ignite. This creates a flame which encompasses the whole lamp. Naturally, this condition can be the source of facial burns and can be disastrous should it occur while descending a drop into a cave. (Chapter 5 recounts my experience with just such an occurrence.)

Since it is necessary to change carbide every hour or so, depending upon the size of the storage chamber and the amount of flame desired, it is important to prevent mud and other cave debris from adhering to the gasket. This can cause acetylene leaks as well.

It is a good idea to carry a number of extra

gaskets in a waterproof container so that if an underground changeover is necessary, it can be done easily. While carbide lamps are not eccentric devices, they contain extremely simple mechanisms and are designed for operation reliability under the most adverse conditions. The only maintenance that is required involves the rubber gasket and the periodic cleaning of the nozzle aperture with a miniature feeler wire which is supplied with most lamps upon purchase. It is, however, also a good idea to carry a spare nozzle or two.

The flame of the carbide lamp is adjusted through the on-top water valve. Usually, this is a set at the halfway point until the lamp is ignited and then adjusted downward for the most intense illumintion. This is produced by a relatively small flame. Larger flames will produce more light, but it is not as directionally intense and tends to make viewing rather blurred. In no instance, however, will the illumination output from a carbide lamp be as focused or directed as that of even the cheapest flashlight. However, it will last far longer in a cave and will put up with a lot more abuse than even the most expensive electrical light source.

Just how long will a single charge of carbide and water last? This depends on several factors, including the size of the two respective holding chambers and the amount of flame you desire. Longer flames require a faster drip and, of course, the calcium carbide is expended more rapidly because it gives off more acetylene gas in a shorter period of time. The size of the chambers in most lamps is pretty much fixed, so you won't find a lot of variation here. I have found that it is often necessary to replace the water before the calcium carbide needs to be replaced. With a slow drip, you might get two hours or more of

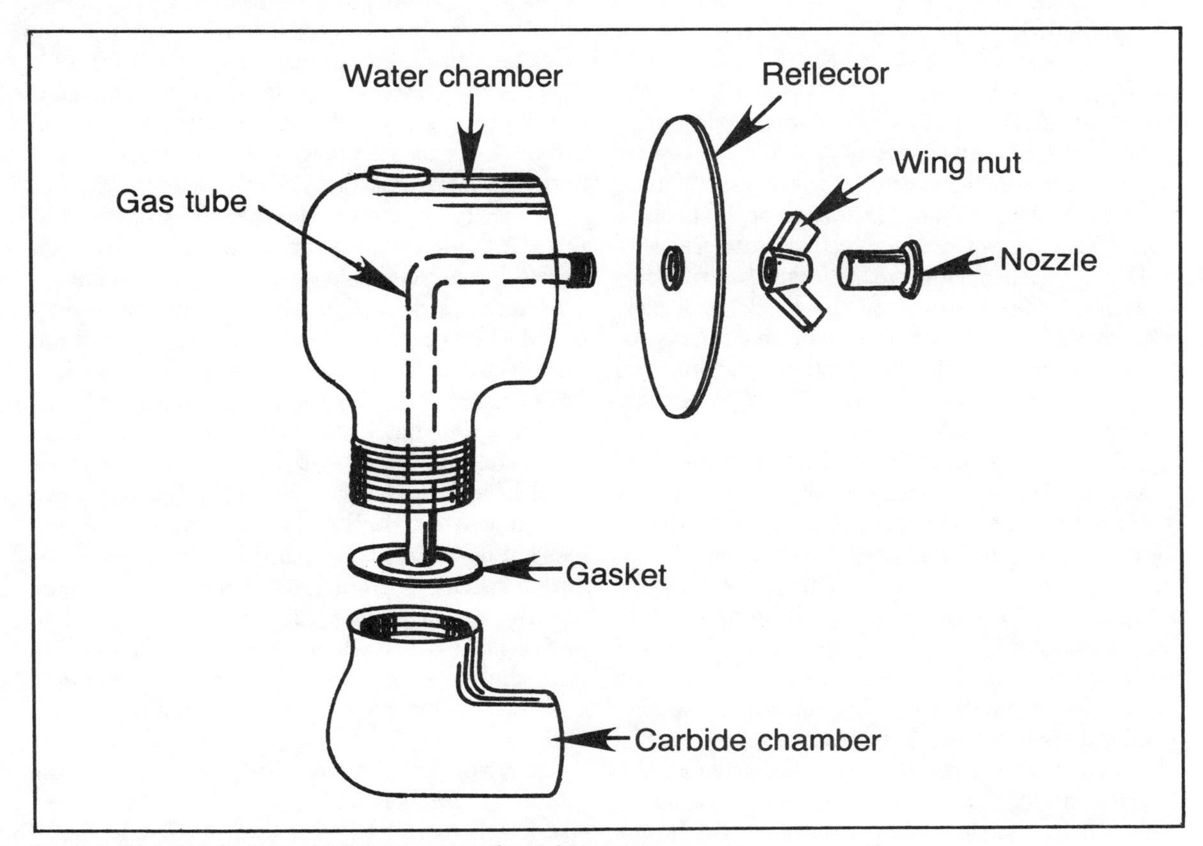

Fig. 4-4. Exploded diagram of the components of a carbide lamp.

flame time; with a fast one, the time of practical use drops to about an hour. When the flame begins to retard, this is a sign that one or both chambers need to be refreshed. Of course, the length of use for any one charge will depend upon how big the charge is. Usually, the water chamber is filled right to the top. You can do the same with the carbide chamber (with calcium carbide, of course), but this can result in a slight problem. As the carbide gives off acetylene gas, it turns into a wet, powdery residue which expands. A chamber which has been filled to the brim with dry carbide often becomes packed with this white powder, choking off the gas from the relatively dry carbide which lies beneath. Also, when all of the carbide is expended, the residue is really packed in there, and it can be quite difficult to remove. For this reason, most spelunkers only fill the carbide chamber no further than the three-quarter point.

Carbide lamps are available in many reflector sizes. Some models allow you to switch reflectors; others are fixed in place. The standard for cave exploring used to be the 4-inch reflector. However, in more modern times, spelunkers have been opting for 1.5 and 2-inch reflectors, which cast a more concentrated light beam. The angle of the beam is not as great, so you can illuminate a wider area with a wider reflector. However, the smaller reflector does not increase the angle too much, and many spelunkers feel the increased lighting intensity for a given area is worth the sacrifice in width. The smaller reflectors are also less likely to become wedged in narrow areas.

My first carbide lamp was the only model available from a particular local sporting goods store. It contained a massive 9-inch reflector which literally hung down below the brim of my hard hat. It produced an amazing light beam but was completely inappropriate for cave exploring. It became an immediate bother upon entering the close confines of my first cave with a group of older explorers who were showing me the ropes. Fortunately, however, just inside the entrance of this first cave, we found a discarded plastic Army helmet which contained a beat-up carbide lamp. The seal had obviously given way and the whole thing had caught fire. It was discarded at this point and probably had been lying there for several years. The old lamp and helmet were completely useless, but the reflector was still in good shape. It was removed in the cave and replaced my behemoth reflector. Fortunately, both lamps were from the same manufacturer, so the reflectors were interchangeable. The problem with the big reflector lay in the fact that it became entangled in every stalactite and overhead ledge encountered. I recommend that you stick with reflectors no larger than four inches. In recent years, however, I have resorted to using a two-inch reflector and I like it even better.

My first lamp contained no instructions whatsoever, so it took a while to figure out just how to ignite it properly. Once the lamp is charged with water and calcium carbide, the drip valve is turned to the halfway point; a few seconds later, you will probably hear a very low hiss as gas exits the nozzle aperture. You can crank on the striker wheel all you want to at this point, but the lamp just won't light. The first time around, I ended up using a match to get the lamp going and couldn't figure out why the striker mechanism wouldn't do the trick. It was certainly producing a magnificent spark.

Shortly thereafter, I learned the proper procedure. As soon as gas begins to escape through the nozzle, place your hand over the entire reflector. This forms a pocket in which a small amount of acetylene gas can collect. After a few seconds, bring the fleshy part of your hand across the striker wheel while pushing downward. This is done quickly and the spark then ignites the gas pocket, resulting in a miniature "pop." That's all there is to it. Of course, my original lamp had a 9-inch reflector, and there was no way to cover it with a human hand (regardless of the size of the human). In this instance, the striker mechanism was next to useless. For more normal-sized reflectors, however, even up to four or five inches, you should have no problem at all. Remember, perform the striking procedure with all due speed. This will prevent you from receiving a slight burn.

Carbide lamp maintenance is fairly simple. The nozzles are cleaned with a feeler wire of the correct size for the particular aperture. If you use a wire

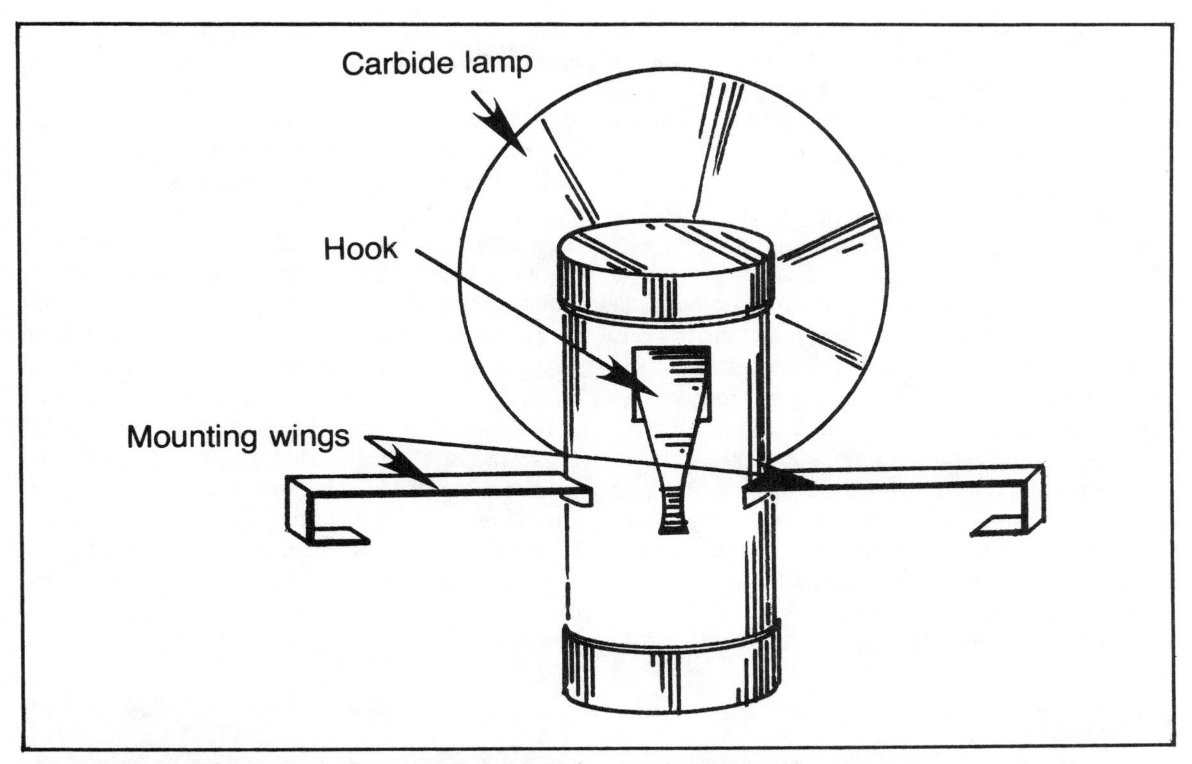

Fig. 4-5. Some carbide lamps are fitted with a small hook for mounting to a hard hat.

which is too broad, it will enlarge the aperture and your lamp will produce a broad, nonconcentrated flame. Carbon builds up from the process of combustion and will eventually interfere with the gas flow. Carbon buildup accelerates rapidly when every last granule of carbide is used to squeeze that last gasp of acetylene through the lamp. For this reason, it's a good idea to carry a feeler wire with you into the cave so that cleaning can be done on the spot. Most nozzles are simply friction-fitted to the gas pipe, so they can be pulled free in a second, cleaned in a few more seconds, and replaced in yet a few more.

Having covered the hat and the carbide lamp, let's discuss the carbide lamp *on* the hat. The main difference in carbide lamps involves their mounting brackets. Some are fitted with a small hook (mounted to the top back portion) and two horizontal wings, as shown in Fig. 4-5. Others use a slotted arrangement which involves slipping the lamp down over a mating wedge, as shown in Fig. 4-6. There are probably many other types out, but most will be similar to these two designs. The latter type is often sold with the mating wedge, a springy piece of steel that is fitted to the front of the hard hat by a nut-and-bolt arrangement. Every installation will require the hat to be drilled at the base of the brim. I used an unusual arrangement, however, to mount the former type; this required no nuts and bolts and the lamp was quickly removable for recharging. First, the front portion of the helmet was drilled with three holes, as shown in Fig. 4-7. The top hole was fitted with an expandable axle pin which formed a tiny eye, through which the lamp hook could be inserted. The metal wings were then bent inward and inserted in the holes far enough for them to expand outward and secure the base. Removal was just a matter of squeezing the wings together again, pulling them free, and removing the hook from the eye. This is the arrangement which is still used today. While the mounting of the lamp to the hat may seem like a minor problem, many arrangements were tried before arriving at this solution. The total cost of the installation as shown will be a few cents for the axle pin.

Whatever mounting arrangement you use, make sure that the lamp will stay on regardless of the situation. All previous mounting procedures I attempted as a teenager were unsatisfactory. The lamp had the nasty habit of partially freeing itself and directing the flame toward my left or right ear, depending upon which improper arrangement we're talking about. A good test is to mount the lamp and then throw the whole arrangement around for a few minutes. Drop it on the ground a few times after a hefty toss in the air. If it stays on, you're probably in business. Obviously, the mounting configuration must also allow for quick removal, as the two will have to be separated in order to recharge the lower chamber. The mounting configuration shown here is certainly not the only one, and this is where you can use your own imagination and judgment to come up with a mount which is best suited for your needs.

Obviously, carbide lamps require fuel to operate. The acetylene gas is generated by the chemical reaction which occurs when water comes in contact with the calcium carbide. For this reason, the storage of carbide when in a cave is most important, especially from a safety standpoint. Caves are usually very wet. If carbide is not protected from the moist environment, it will react and a small amount of gas will constantly be given off. (We're talking here about the carbide that is to be used for recharging the lamp, not that which is already packed in the bottom chamber.) Calcium carbide is sold in cans which are usually fitted with watertight plastic tops. The can is vacuum-packed at the factory, but once opened, make sure that plastic cap is in place. Depending upon the size of the can, it may be difficult to take the entire amount with you on most expeditions. Even if you plan to stay for quite a few hours, you won't need much more than a couple of

Fig. 4-6. Other carbide lamps use this slotted arrangement.

Fig. 4-7. To attach my carbide lamp, I drilled three holes through which the metal wings could be slipped.

handfuls to keep your way adequately lit. Cans are rather bulky, so it may be best to remove an adequate amount and stow it in a waterproof container. Plastic sandwich bags work nicely, although it's best to use an arrangement of several bags; one inside the other, to make sure that the package is absolutely watertight. Remember, you'll often be crawling through highly saturated mud and even through some small streams, so whatever packing arrangement you use should have the ability to keep all water out, even when totally immersed. I have also used Tupperware™ food containers for this purpose. Any type of flexible waterproof container will work as well. Make certain that the lid is snapped on tightly and always check for any signs of puncture or tears which can allow water to enter. In the most atmosphere of a cave, the air alone can cause a chemical reaction. This is not unually dangerous, as the gas given off is not under pressure, but one never likes to take the chance of breathing toxic and/or flammable gases while in the uncertain confines of a cave.

ELECTRIC LIGHTS

While carbide lamps are almost universally used for cave exploring, over the years some spelunkers have come out advocating sealed-beam, battery-operated illumination. Several types of lights are available that contain the bulb and reflector in a separate unit which attaches directly to the hard hat. The batteries are usually mounted in a small case which is clipped to the spelunker's belt. A two-conductor flexible cable connects the two units. These lights are more efficient than carbide lamps as far as illumination is concerned. The focused beam provides more definition, although the beam width is not usually as large as a carbide lamp's. Generally, these electrical devices are rugged but not as rugged as a carbide lamp, which contains no glass elements whatsoever. Spare battery packs can always be carried, but make absolutely certain that they contain full charges before entering the cave. There's always a possibility that the light bulb will burn out, so spares should also be taken along.

If you elect to use the electrical illumination, make certain that the lens or cover placed on the electric light is of the unbreakable variety. It's almost certain that it will be banged up against the cave wall or some overhead formation during any exploration. Clear plastic lenses are quite acceptable, but you will find that they will become badly scratched after only a few outings and replacement will be necessary. Also, the battery compartments should be sealed against moisture and mud. When water infiltrates an electrical circuit a current drain is often placed on the battery, which can quickly discharge as a result. Also, the connecting cable between the lamp and the battery pack is prone to snag on every projection you come in contact with. Often it's necessary to put coveralls over the battery pack and cable to keep this from happening, but the cable still exits near the collar and can get snagged up at head level. Then too, should your hat fall from your head, there is a chance that the cable may break, causing the immediate cessation of illumination.

I have used electrical hard-hat-mounted lights on just a few occasions and I don't really like them for general cave exploring. However, when exploring large caves that are known to have wide, evenly negotiated passageways, they are often handy to have along *in addition to* the carbide lamp. When exploring these large passageways, the carbide lamp can be unclipped from the hat and replaced with the sealed beam unit for better illumination. Some say they prefer electric lights when doing any climbing in a cave. There is some concern on their part that when using a carbide lamp there is always the possibility that the acetylene flame may come in contact with the rope, causing it to burn and weaken. While I suppose that is a *possibility*, the danger is so slight as to be corrected by simply keeping the carbide lamp reflector away from the rope. In the first place, ropes are generally very damp after only a few minutes of use in a cave, so to cause any real damage, the flame from the carbide lamp would have to be kept in close contact with the rope for a very long period of time. I feel that the danger of hooking the electric cable around a stalactite or other prejection during a descent and possibly losing all light is far more dangerous. Therefore, the carbide lamp is always used when I am doing any climbing.

As a teenager, I designed and built an electric hard-hat-mounted light which suffered from fewer of the problems associated with this type of illumination when used for cave exploring. The device was made from an old flashlight and used the lens, bulb, and reflector assembly mounted directly to the brim. There was enough space between the hat webbing and the inside top portion to accommodate two D cell batteries in a plastic holder. The holder was attached to the inside top of the helmet with epoxy cement and a short cable was run (again, inside the helmet) to the bulb, routed through a small hole drilled in the front of the hat. This provided electric illumination and contained no external battery pack or cable. The whole assembly was self-contained.

This arrangement worked fairly well; the beam was a bit better than that which could be obtained with a conventional carbide lamp. However, the reflector became severely bent after the first outing. It was replaced with another unit more rugged than the first, and it was used several more times without incident. After a few months of this, however, another problem made itself known. Due to the moistness of most caves, corrosion set in and there was a buildup of insulation between the bulb and its holder. The light became erratic and would dim or go out completely, usually at the most inopportune moment. The assembly required an extremely thorough cleaning after every outing and I finally gave up on it completely in favor of the more dependable operation of the time-proven carbide lamp.

It is easy to pick up similar types of electric lights at most department stores. Some of these may be all right for hard hat mounting, but I recommend that they be taken along as a *fourth* source of lighting. Choose the type that is made of molded plastic and contains a fairly small battery pack. These can be used for closer examination once you're in the cavern and may be fitted with appropriate straps so that they can be temporarily attached to a hard hat in order to free your hands. Do not allow these, however, to replace the common flashlight, at least one of which should always be included in the spelunker's emergency lighting complement. Flashlights should be chosen for construction. Make sure the unit you select is completely sealed. You may be able to find very rugged types at war surplus outlets, many of which are designed to operate while completely immersed in water. Some spelunkers choose two-foot long behemoths that produce a good light. Those, however, are generally too large and bulky for practical use in caves and will consume far more packing space than their purpose deserves. The standard two-cell (D type) variety is generally best, and make sure you also include at least one extra bulb and one fresh complement of batteries for each light you carry.

EMERGENCY LIGHTS

It's more important to pay attention to your emergency lighting sources and accessories than any other equipment you may carry. Unfortunately, emergency supplies are often purchased and then neglected or even forgotten because they are so seldom used. An emergency flashlight is worse than no good if you've allowed it to remain in a damp knapsack for months or even years, as it may fail you at a time when you thought you could depend upon it. Before entering any cave, all emergency lighting sources must be checked to make certain they are intact and fully operational. If they are not, the exploration should be scrubbed until proper repairs can be made. Make certain that the top portion of your flashlight is screwed tightly in place. Even waterproof models will allow moisture to infiltrate when they are not properly sealed. A broken lens will allow the same type of electrical corruption. Be highly suspect of any flashlight that has received an intense mechanical shock from having been dropped. The damage may not be obviously apparent, and a close examination is mandatory. If there is any question as to whether or not the unit is still moisture-proof, then it should be replaced and relegated to aboveground use only. Even long-life batteries will fail after a period of time, and even if never used. For this reason, periodically remove them from the flashlight and check for signs of corrosion, leakage, or water infiltration. Once water has been allowed to enter your flashlight, even for a short period of time, it should be replaced with a new unit. Remember, this is emergency lighting, and if it is ever needed, it *must* be functional and fully so.

The next emergency lighting source is usually candles. In over twenty years of cave exploring, I have been forced to resort to my emergency flashlight only twice, but never to candles. The candles are there in case the flashlight fails (and assuming that the carbide lamp has also failed). As an experiment during one expedition in a rather large cave, one group I was with extinguished our carbide lamps and resorted to candles. This was a simulation of an emergency situation which assumed that all sources of light had failed save for the candles. The going was difficult, to say the least. Large caves can be rather drafty, and the candle flames were constantly being blown out. The minimal light produced forced us to crawl slowly on hands and knees in order to see the floor properly. The going was difficult because obviously one hand had to hold the candle. Incidentally, the candles burned very quickly due to the strong drafts and didn't last nearly as long as one might imagine. After twenty minutes of proceeding like this, we unanimously decided the experiment was over and gratefully popped the flames of our carbide lamps back to life. We determined that yes, we *could have* safely and successfully exited the cave with the very dim illumination of our candles alone, but the rate of progress would have been cut by about a factor of five. Remember, too, that this was a fairly open cave with wide, level passageways. The difficulty factor of using candles to exit a tight, winding cave with many obstructions might be increased ten times. This is mentioned to point out just how much you must depend upon your carbide lamp, and flashlight should the former fail. When you must exit a cave under candle power, you really are in an emergency situation, one that should never have occurred in the first place. The illumination given off by the burning wick is better than no lights at all . . . but not much.

While I have not used them in cave exploring, there is a product available today which contains a phial of a phosphorescent chemical which will provide a greenish glow for an hour or so. The flexible phial is twisted, which allows chemicals from two compartments to combine and produce the glow. It would seem reasonable to expect these devices to serve a useful purpose as emergency lighting sources for spelunkers. You certainly wouldn't be plagued with a flame that is constantly being blown out or hot wax dripping on your hands, as is the case when using candles.

CLOTHING

Cave clothing is discussed throughout this book, but a special section is devoted to it here because the clothing you wear while spelunking can be considered a very vital part of your equipment. Starting at the bottom and working up, the shoes or boots that are worn can make the difference between a safe exploration and one that is extremely risky. For example, leather-soled shoes become very slippery when just a small amount of moisture is present. In a cave, you are dealing with excessive amounts of moisture, and a few seconds after entering you will be able to get very little traction if this type of footwear is worn.

Any sporting goods store can supply you with good footwear of the type designed for activities where traction is of utmost importance. Hikers' boots work very well, but these are sometimes difficult to obtain in full waterproof designs. While it is not necessary to buy your footwear with full waterproofing in mind, this can be a decided comfort advantage, especially during long expeditions. I prefer lightweight, mid-length hiking boots for explorations which do not require wading through moderate-sized streams. For the latter, a pair of calf-length rubber boots is usually donned. These are relatively inexpensive and can be stowed in a very small compartment of a knapsack if thin rubber types are chosen. Of course, these will not provide a great deal of traction, which is necessary when wading through swift running water. Heavier one-piece rubber boots which are molded and contain thick ridges on the soles are best for this application, although they are not easy to store in a knapsack and can be rather heavy.

Many amateur spelunkers wear tennis shoes for all explorations. These are ideal for climbing situations and generally provide reasonable traction, but they are anything but waterproof and can make for difficult going down passageways which contain a moderate amount of small rubble. Tennis shoes provide very little protection from the certain bumps to the feet which will be encountered in every expedition. It's much easier to stub a toe. On the positive side, however, this type of footwear is inexpensive and can be washed and even replaced at frequent intervals. As ideal caving footwear, however, they leave much to be desired in the overall picture.

In choosing your footwear, stay away from most types of hunting boots; these tend to be rather heavy. Weight is a major problem in any expedition, and if it's contained at a low center of gravity on the human body, as is the case with all footwear, the problems are multiplied. A spelunker who has to drag heavy boots will tire very quickly, and frequent rest stops will be required. The ideal tradeoff is a lightweight hiking or climbing boot which contains some water-resistant properties and is frequently treated with a waterproofing compound (also available at your local sporting goods store).

More important than the footwear itself is the fit. Nothing slows or even ruins an expedition quicker than a boot which is too tight (cramping the foot) or too loose (allowing the foot to slide and blister quickly). I have extremely narrow feet and often find it difficult to obtain a size 11 boot with a B width. These can be had on special order but tend to be very costly—and caves just love to eat up footwear. Worse yet, I also have flat feet, so it is necessary for me to pay particular attention to all footwear purchases. The flat feet problem can be easily overcome with an inexpensive pair of good arch supports. If a boot is slightly wide, this can be taken care of by wearing an extra pair of socks. You want to be careful here to avoid *too* much insulation around your feet causing perspiration and an eventual arthritic cold feeling throughout your lower extremities. Uninsulated, lightweight footgear allows your feet to breathe and cuts down on this possible complication.

Only lace-up boots should be considered for cave exploring. This precludes the rubber type which should be in one-piece form, containing no snaps, hooks or zippers of any kind. The latter will quickly become clogged with mud and will be very difficult (if not impossible) to remove without some footwear surgery. Lace-up boots are far easier to manage but should be tied very firmly (with a square knot) to make certain they do not become undone at a critical point in the exploration.

Break in your shoes well ahead of the expedition. Nothing is quite so bad as a stiff pair of boots which feel comfortable at first but quickly take their toll after the expedition is underway. When the painful breaking-in process has been completed, you will find that the increased comfort was well worth the effort.

Footwear care is most important and should be preferred immediately after an expedition. This involves removal of all mud and other debris, a thorough washing or wiping with a wet cloth, and immediate drying. If the soles of your boots are fitted with traction ridges, pry out any small stones or twigs that have become embedded in the slots. While caves are very rough on footwear, maximum adherence to this last detail will make them last a long time.

Traditional garb for spelunkers includes a pair of coveralls. These should be of lightweight material, of the button-up type (rather than zippered), and preferably fitted with long sleeves. The one-piece design will go a long way toward removing the possibility of your clothing becoming entangled in the many thin projections that you will be crawling around and over. Many amateur spelunkers prefer blue jeans and flannel shirts. These are all right but are best worn beneath close-fitting coveralls. Naturally, you do not want to dress *too* warmly for the expedition, because the possibility of perspiration becomes more likely. In most instances, you should feel just slightly cool five minutes or so after entering the cave and before any strenuous exercise has been necessary. Caving does involve moderate to short bursts of strenuous activity, so if you're snugly warm before any of this starts, you will probably be sweltering hot when the work really begins.

Rubber knee pads are often used by spelunkers. These can be purchased at a local hardware store and are fitted to the knee with an elastic band. Knees take an awful lot of abuse when crawling on a hard stone floor, so consider some sort of protection for this area of your body. In lieu of knee pads you can substitute additional layers of cloth sewn over the knee areas of your coveralls.

While it may seem a bit ridiculous to discuss underwear in a book on cave exploring, this again is part of your equipment complement and can have a great effect on comfort during an expedition. Many spelunkers wear long underwear, especially if they intend to be in a cave for an extended period of time. Thermal underwear is lightweight and is designed with pores that allow the skin to breathe and rid itself of moisture. One or two-piece designs may be used, since your outer clothing and coveralls will provide an external one-piece design. More conventional types of underwear are fine for the great majority of expeditions which do not entail being in a cave for a long period of time. Briefs or boxer shorts should be loose-fitting to avoid binding. Remember, your entire body is going to be twisted and turned into a number of unusual positions, so clothing flexibility is very important. Male spelunkers may also wish to consider a high-quality athletic protector to avoid possible injuries to the groin area.

For most spelunkers, gloves are optional. They can be very useful during certain parts of an expedition and a complete nuisance at other times. During any rope-climbing phase of an exploration, I always remove my gloves for the best possible traction and feeling on my lifeline. Others prefer to wear close-fitting gloves for this type of procedure, the type which has been made with a rough palm and finger surface for better traction. These do provide increased traction capabilities when compared with the human hand, but I stay away from them; when I'm climbing, I want to be able to feel the rope as much as possible. Some spelunkers say that gloves decrease palm and finger fatigue. While this is true concerning abrasions and blistering, gloves do nothing to actually delay fatigue, which is mostly concerned with the muscles and joints.

In any event, gloves used for climbing should be thin and allow for maximum finger movement and manipulation. Gloves used for general exploration can be of almost any type, although suede or leather weatherproof types will probably give you the most service. Avoid gloves that are extremely bulky. These will decrease your hand sensitivity to the various objects you will be depending on to pull you along a passageway and help you over small rises. Gloves will also prevent the many cuts and scratches on a spelunker's hands from sharp rocks, rough cave walls, and other underground obstacles which will be encountered almost continuously.

ROPES

Every spelunker will need ropes at one time or another. Generally speaking, you want a high-quality rope of sufficient width to provide a firm grasp. I have often used half-inch manila, which is a bit small for my liking, but a large amount of it can be carried without an undue weight problem. For serious climbing, the three-quarter inch size is preferable, and for pure vertical descents of quite some distance, the one-inch size is even better. Large-diameter rope is quite heavy, so it is most often used in caves which have been previously mapped and are known to contain a sheer drop or two. It is only necessary to carry the large rope to the drop, where it is fixed and left in place for the climb back up again.

Regardless of the size of rope used, the main thing is its condition. When descending potentially lethal drops, a half-inch safety line is always used and is tied around the waist of the climber. After each expedition, the safety rope should be discarded and replaced with a new one. This may sound quite extravagant, but when you need that safety rope, you really *need* it. It is there to stop your fall should you slip from the main climbing rope or if it should break for any reason. I once witnessed the beginning of an expedition by a cave club from Pennsylvania, the members of which were depending upon a one-inch manila rope to descend into a cave opening that began with an 80-foot drop. I was to explore the same cave the following week and was looking over their equipment. This was supposed to have been a quite professional outfit, and I was very surprised to see that they used no safety ropes, and that their main one contained a break in one of the four intertwined strands. When I pointed this out, the leader told me that the rope was plenty strong even with the break and that it had served them well for several years. They made it in and out all right, but they were taking a tremendous chance with their lives, and I lost all respect for that particular organization. For goodness sake, if you can't afford an intact main rope and an inexpensive safety line, then you're just not well off enough financially to even consider cave exploring.

Some irresponsible spelunkers have been known to leave ropes in caves. These are often old ropes that were about to be discarded anyway and which may have become entangled in a rock or other obstacle at the bottom of a drop after an ascent was made. Instead of climbing back down again and freeing the bottom end of the rope, these parties simply left it in place. As rough as a cave environment is on a length of rope, just leave it there for a few weeks or months and it becomes very weak. *Never* depend on a rope or any other piece of cave equipment that has been left underground for an undetermined period of time. If you find a length of rope in a cave you are exploring, by all means take it out with you and throw it in the nearest trash receptable. The entire caving community will be better off for your efforts.

OTHER ACCESSORIES

So far this discussion has covered only the basic equipment involved in cave exploration. There are, however, a large number of specialized tools and devices that may be needed from time to time. One very handy device is a collapsible ladder which consists of two parallel sections of rope which have been fitted with metal (or sometimes rope) rungs. More modern types are made entirely of metal and use two flexible lengths of metal cable to replace the vertical sections of rope. Lightweight aluminum rungs are fitted every foot and a half or so. Such ladders make descents down sheer drops much easier. This is especially true when the drop begins with an overhang and would normally necessitate a hand-over-hand or rappel technique with a standard rope. Hand-over-hand is tremendously dangerous, and rapelling takes special knowledge of climbing. On the other hand, with the ladder, you simply attach it at the top of the drop and let it fall vertically to the cave floor below. You then descend, one rung at a time.

Whenever a descent is made with a ladder, a safety line must always be attached to the climber. Never depend upon the ladder alone, nor your ability to stay on it, to keep you safe. The safety line is tied around the midsection of the individual and is played out as the descent is made. When at the bottom, the spelunker removes his safety line, which is then hauled to the top again for someone else's descent.

Techniques used for descending into caves using ladders differ, depending upon whether the ladder is made of cable or rope. Using the cable ladder, one foot is placed on the first rung, and the other foot is placed *around* the ladder on the next rung. In other words, your left foot will be on one side of the ladder while your right foot will be on the other side. You do not climb or descend in the same fashion as you would a standard stepladder, going up or down the rungs on one side only. With the cable ladder, the spelunker is actually straddling the left or right cable. With a rope ladder, the arms are used mainly for support, with one hand grasping the vertical portion of one suspension rope from one side and the other holding the one on the opposite side.

While collapsible ladders are often handy, they are not often carried as standard gear by most amateur cave explorers. These devices can be rather expensive, especially where long lengths are involved. For this reason, they are usually incorporated only when a known situation exists in a previously explored cave that dictates their use.

Some caves have large, deep streams or lakes which must be forged or navigated for complete exploration. Here it often becomes necessary to utilize some type of flotation device. Early spelunkers often resorted to an inflated automobile inner tube, but of course this meant getting drenched to the skin. In more recent years, inflatable rubber rafts have become popular. Most often the one-man raft is used, each spelunker carrying his own. In recent years these simple rafts have become rather inexpensive; one can probably be purchased for less than $50. Some are inflated with air while others use a special inflation system which generates gas through a chemical reaction. The latter is to be avoided. Most one-man rafts can be inflated in a short time with a small air pump and deflated in a similar period of time. If you have need of an underground raft, choose the type which is fully collapsible, containing no wooden seat or bottom. Also, make sure the device is made of a tough, rugged material, as you will be quite frequently rubbing against cave walls and other sharp obstructions.

As was the case with the rope or cable ladder, rafts are highly specialized equipment and are not generally included for most explorations. Caves with large underground rivers and/or lakes are quite rare, so most of the time a raft will not be needed. Then too, many underground streams are not navigable, in that the cave walls surrounding the body of water are too close together, and often the ceiling will come down to within a few inches of the water's surface. Exploration from this point on requires that the spelunker lay on his back, keeping his nose high, and push himself under the ledge until the passageway increases in height.

There are many other specialized pieces of equipment that will either come in handy or even be mandatory in exploring some caves. Caverns which require a great deal of ropework will also require the use of professional mountaineering equipment, including pitons, piton hammers, caribiners, etc. At this point, however, we are approaching the professional side of cave exploring. Such pursuits are not often within the realm nor even the desire of weekend spelunkers.

Photographic equipment forms another specialized area. Special techniques are often required for use in a cave, although you might be surprised at what you can do with an inexpensive camera and a flashcube or two. Perhaps the most important photographic equipment includes the case used to transport the camera and its accessories. I use rather expensive equipment when doing photographic work in a cave, so I had a special aluminum camera case built which will also hold extra film, a flash attachment, and so on. It is fitted with foam rubber padding which protects the equipment from severe shocks and also keeps it firmly in place. Additionally, the case is completely waterproof and will float indefinitely if accidentally dropped into a body of water.

If you're interested in underground photography, it will probably not be necessary to go to the expense of purchasing an expensive waterproof case. You can probably pick up a war surplus ammunition box for a few dollars. This will give you the waterproofing that you need, but it will be necessary to buy foam rubber padding, cut it to size, and glue it in place. Padding must be installed on the bottom, both sides, and the top. A good test of the shock-resistant ability of the padding is to place an egg inside the box and then throw it up in the air a couple of times, allowing it to strike the ground. If the egg breaks you have a terrible mess to clean up, but this is a certain indication that more padding is needed. Remember, there is always the chance that your hand will slip from the handle of the box, and it, along with your camera equipment, can fall many feet. Most cameras today are very rugged, but they are not up to the rigors of spelunking without some additional protection.

A first aid kit is also a mandatory part of every spelunker's equipment. Any nicks or cuts incurred during an expedition should be attended to immediately to prevent infection from the bacteria-rich cave dirt. A first aid kit need not be large, but it should be well-stocked with adhesive bandages, disinfectant, and most of the other minor medical supplies which you keep around the house. The subject of first aid kits and their uses is more fully discussed in a later chapter on safety, but how these supplies are stored is another matter unto itself. Many first aid kits that are purchased commercially are not all that waterproof or shock-resistant. For many years I have carried a personal first aid kit which consists of a small vial of antiseptic, numerous bandages, and even a tiny snake bite kit all housed in a plastic container originally designed to hold two packs of cigarettes. The container comes in two sections, one telescoping into the other. It is waterproof and weighs only a few ounces fully stocked.

Whatever type of container you use, it should be completely waterproof, able to keep out all cave dirt, and of high-impact design. I recommend each person in an expedition carry his or her own miniature kit, but stock it will enough that it can take care of the needs of additional persons should the need arise. Again, the complement need not be tremendously wide in scope. Stick with the essentials: bandages, antiseptic, adhesive tape, smelling salts, etc.

Incidentally, the snake bite kit mentioned previously is carried because to get to many cave entrances it is necessary to wade through a great deal of underbrush. The danger of snakes really only exists aboveground and possibly in the entrances of caves. Generally speaking, snakes are not seen in the interiors of caves, although the odd exception can always exist. I have never seen a snake in any cave I explored, although in the western part of the United States snakes may be more prevalent in entrances, where they travel to get out of the hot sun. Be cautious, however, in the entrances of *all* caves, especially those which start as a vertical drop. It is possible for a snake to fall through the opening and lay for quite some time on the cave floor before dying of starvation. It is a good idea to check with local residents when exploring in distant areas to find out what types of poisonous snakes (if any) are common to the area. Again, I have never seen a snake (alive or dead) in any cave I ever explored, nor have I heard of any local spelunkers encountering them. Rumors do run rampant about caves in the West which are literally filled with rattlesnakes, but to the best of my knowledge, these incidents have never been documented.

Although done very rarely, some expeditions find it necessary to set up some form of communications equipment with the surface. We will not deal with this subject in great detail; again, the need for communications equipment is extremely negligible. First, don't think you can get away with a simple walkie-talkie. In most instances the signal will be completely blocked by your underground enclosure. Most communications equipment involves the use of wire connecting lines between the surface and the underground, and war surplus field telephones are quite common. These are rugged devices that can be picked up inexpensively from war surplus outlets. Again, communications equipment is used to talk to individuals who remain on the surface. This provides an emergency system in some cases and may be applicable to expeditions which plan to remain underground for several days. It is not necessary to have this equipment for communications among members of the expedition who are underground, because the group is supposed to stay together. The last thing you want is small segments of the expedition going off on their own.

CAMPING GEAR

Even if you plan to explore a cave for only a few hours, you will need a fair amount of equipment generally associated with hiking and camping in order to take care of your needs. A fresh water supply must be carried by each individual. This is most often contained in a canteen strapped to the belt, or better yet, kept inside a small knapsack. While there is a great deal of expensive camping gear available at your local sporting goods outlet, you may find that you're better off at a war surplus store. While metal canteens are very rugged, I prefer a molded flexible plastic canteen, which I have carried for over ten years. This device weighs very little and has lasted longer than any metal canteen I have ever owned. Metal canteens quickly show the stresses of cave exploring, but a flexible plastic canteen will absorb many bumps and accidental drops without any sign of damage. They are often less expensive than the metal type and don't seem to be as prone to leaks, since there are no seams. I keep my canteen in a knapsack, although many prefer to fit them with a belt mount and carry them around the waist. However, even when worn below coveralls, this extra bulge can easily become entangled in a rocky projection and hang you up momentarily. You might wish to try both methods of transport in order to determine which is best suited to your own personal spelunking needs.

As far as the knapsack is concerned, war surplus is still the best way to go. You can pick up a very rugged knapsack from such an outlet that will probably last you all your exploring days. Whichever route you go, choose a knapsack that is compact. The top and sides should not extend outward or above the human body. This prevents hangups in tight passageways. The separate compartments can be used for stowing your canteen, medical kit, and any other gear you may wish to carry. It's a good idea to treat all knapsacks with a good waterproofing compound that will keep its contents reasonably dry, even in extremely wet conditions. Of course, all perishable items held within the knapsack should be contained in their own separate watertight containers.

Other items which should also be taken into a cave include a pocketknife, a good supply of matches (in a waterproof container), and possibly a collapsible cookstove. The latter can be used for brewing coffee and cooking hot meals in well-ventilated caves when the expedition length warrants this type of procedure. The pocketknife will come in handy for cutting away jagged rope ends and in a number of other applications. Make sure you choose a folding pocketknife. Do not go into a cave with a hunting knife strapped to your side. Even in its protective leather case, it can easily become dislodged and create a dangerous situation.

Long expeditions may require you to carry food with you, but even if you're involved in a very short one, carry at least a few edible items that can give you an energy boost and possibly sustain you should a minor emergency occur which requires a longer stay than originally intended. Chocolate candy bars are excellent energy boosters; this high-caloric food has sustained individuals for days. I always take along a small supply of canned goods in a special compartment of my knapsack. These foods have never been used, as they are intended for emergency purposes only, but they're there if I need them. They are packed in cans with lids that are removable by hand, so no can opener is necessary. For some long stays in caves, spelunkers often carry a shockproof Thermos or two to carry coffee, hot chocolate, soup, or other warm liquids. After you've been in a cave for a few hours, the humidity can really get to you, and a warm beverage can give you a whole new outlook for the remainder of the expedition. Incidentally, a wide-mouth Thermos can also be used to store other hot food items of a more solid nature. While not specifically intended for this purpose, I have found that you can cram a lot of hot meals, from asparagus to steak (cut up in small pieces) into even a small Thermos, although it may take a long-handled spoon to get your meal out of it. Remember, use only a rugged, unbreakable Thermos. The glass-lined types have a life expectancy of about two seconds in an underground environment.

If you plan to eat meals underground, you may wish to visit your war surplus outlet again and pick up a GI mess kit, a compact package that consists of a small skillet, plate and skillet lid/saucepan, all contained in one unit. These may also contain a complex gadget which serves as a knife, fork, spoon, and can opener, all rolled into one. The entire kit is about eight inches wide and four inches high and supplies you with most of the comforts of home (if you improvise a little). This type of equipment is necessary only when an extended stay is planned that will definitely require you to eat several meals underground. While cooking implements and a means of keeping food hot are useful in some applications, most cave explorers simply pack a sandwich or two and a canned soft drink wrapped in tin foil. While not warm, any meal underground will generally taste pretty good and give you the added energy needed at a time when you're beginning to tire.

Overnight expeditions require a good waterproof sleeping bag which is compact enough to be rolled into a tight package. You should also carry an inflatable air mattress, as sleeping on a cave floor is almost impossible without one. While the temperatures of most caves are not especially cold (in the mid-fifties), the humidity must be considered. Due to the combination of the temperature and humidity, an overnight stay can chill you to the bone if your sleeping bag does not contain adequate insulation. You will certainly be sleeping in your clothes, which will be fairly dry if you have worn a good waterproof coverall, which is removed before retiring. Any reasonable quality sleeping bag should do the trick. The air mattress keeps you comfortable in two ways—first, it prevents your back from contact with the hard and uneven cave floor. Secondly, it insulates you from the coolness of the floor and keeps you warmer. Chapter 5 describes in detail an overnight stay in a cave, with attention to the many odds and ends needed for such an expedition.

In summary, there are several essential items which must be carried with you anytime you enter a cave. All equipment and accessories must be checked ahead of time to make certain that everything is operational. Any piece of equipment which is not functioning properly can lead to more problems than not having it along at all. When you don't bring something, you don't depend on it. When you do, you expect it to be operational when you need it. The specialized equipment mentioned in this chapter is used during some expeditions when conditions are known. This would apply to caves which have previously been explored, often mapped, and this information is garnered ahead of time. It's usually ridiculous to enter most caves with a full complement of rubber rafts, cable ladders, communications equipment, and so forth. Chances are you just won't need them in most instances. However, if you ever have the opportunity to enter a large cave in which exploration conditions are unknown, this may be necessary. In such instances a smart group does preliminary exploring ahead of the main outing so there is an element of anticipation involved, and an educated guess can be made as to the gear that will be needed.

Cave exploring equipment need not be expensive and should not be complex. Many items can be made at home by modifying more common aboveground implements. Those that must be purchased can be had for a reasonable price, so your entire complement of gear should cost you less than $100—possibly less than $50 if you are resourceful.

Chapter 5

Exploring a Wild Cave

The most exciting aspect of cave exploring is realized when it comes time to enter your first wild cave. There is an element of the unknown, which is certainly important, but to add to this excitement is the preparation that *must* be done, often days or weeks before the actual expedition is to start. Now, in most areas you will not have the opportunity to enter a previously unexplored cave or cavern system. Although spelunkers are few in number, they do get around, so most caves have been explored time and again and probably for the last several hundred years at least. There may be rare exceptions to this. Sometimes, previously unknown caves are uncovered due to excavation, and some people actively seek out aboveground limestone formations, looking for signs that a cave may lie buried beneath. These signs include the presence of tiny openings at ground level, the discovery and camel crickets on limestone layers aboveground, etc. If there is an indication that a cave lies below, pick and shovels are brought in to uncover an opening large enough to crawl through.

Many such exercises end with absolutely nothing. A few others may uncover the entrance to a small cave or fissure in the limestone. Still others may lead to the discovery of a vast cavern system. This was the case in Front Royal, Virginia, when Skyline Caverns were first discovered. The late Dr. Walter S. Amos, during the 1930s, suspected the presence of a large cavern system somewhere in this area, which already contained 2500 known caves. He began a diligent search to discover a large cave by walking the mountains and examining aboveground limestone formations. At this time the only major entrance to Skyline Caverns was completely covered by earth. Dr. Amos had noticed that some small streams in one vicinity seemed to disappear underground; he finally discovered a limestone ledge just beneath which camel crickets could be seen entering and exiting. He and his party began excavation at the base of this ledge and several hours later uncovered a small stalactite formation which had been partially severed. Following the sloping of the limestone ledge, he and his party finally broke through to the actual cavern system.

The following day, this group of spelunkers

entered the caverns and discovered a comprehensive network of passages filled with stalactites, stalagmites, flowstone, and other beautiful formations. Later the caverns were opened to the public and still serve as a major attraction to the Front Royal area over fifty years later.

Dr. Amos and his party not only discovered a previously unknown (and apparently unexplored) cavern system, but also a type of calcite formation whose existence was unknown in any other cave in the world. He named them anthodites, which is a Greek derivative that interprets as "cave flower" in modern English. Skyline Caverns still advertises that they are the only caverns in the world to contain anthodites.

Few spelunkers in this day and age will be fortunate enough to make such a discovery and enter a cave that has not been explored since prehistoric times. However, the joy and excitement of entering any wild cave is an experience which makes all the forethought and planning worthwhile.

Previous chapters in this book have laid the groundwork for cave exploring. In this chapter I will recount some of my experiences (and adventures) in exploring wild caves. To date, I have explored more than 3000 caves and cavern systems and have found that no two caves and no two expeditions have been exactly alike. You may not experience the multitude of happenings that I have, but when you enter your first wild cave, you may see some similarities in your experiences and mine.

Before the stories begin, I should point out that the names of caves used here are real. However, most caves contain no names whatsoever, and the identifiers given for the various caves explored are those that were used by local residents and some that I made up later.

SINK-HOLE CAVE

When I first became interested in cave exploring, I enlisted some friends who were equally inclined to go underground for a few hours each weekend. Sink-Hole Cave was a large depression located a few miles from Skyline Caverns just off the highway. It had been rumored that the original owner of the property on which the entrance was situated had attempted to develop Sink-Hole Cave at one time, but gave up due to the costs involved. I had seen the opening many times, which in itself was quite impressive. It was circular, entered the ground at a 45 degree angle, and was a full 30 feet in diameter. The opening alone promised a tremendous cave beyond the shadowy reaches which could be seen from outside.

Three of us met at 7:00 on a Saturday morning in the spring. We had heard rumors that some of the passageways became partially flooded due to heavy rains (and to spring thaws), so we chose a time during a dry spell. Rain had not fallen in the area for over two weeks, and the weather had stabilized to an average of 65°.

We had heard that a cave exploring club from Pennsylvania had entered the cave some years before but could not get a line on the persons involved, so no information could be obtained as to what to expect. Anticipating just about every situation imaginable, we proceeded through the entrance at 9:30 that morning equipped with three sources of light (carbide lamps, candles, and flashlights), 250 feet of rope, and even a rubber raft which one of us had purchased at a surplus store. I should also point out at this time that the age of all three explorers was seventeen.

A lot of preplanning had been done for our first outing. Knapsacks were filled (but not overly so) with spare carbide, canteens of water, camera equipment, food (for an in-cave picnic) and even a first aid kit. Parents were notified as to where we were, what time we should be back, and what time to press the panic button. If the truth were known, my mother had already hit the panic button several days before, but was subsequently calmed by my father who had done his own fair share of frightening parents by the same antics forty years before.

All the reading that had been done about caves and cave exploring, all the preparation and planning and all the intense desire to go underground did absolutely nothing to assuage the fear and trepidation that comes over you when you begin the *slow* entrance into the mouth of your first wild cave. Every horror flick that I'd ever had the misfortune to suffer through came to mind, even though I knew that the chances of Godzilla lurking in a cave in some small Virginia community were probably less than 10 percent. Arguments over who would be president of our three-man cave exploring club ended in seconds when I was suddenly voted in two to one (with me casting the dissenting vote) to the top spot and then told to lead the expedition. With heart in throat, I walked through the threshold and my whole world changed. The fear left me as quickly as it sprang up once I could see (by the glow of the carbide lamp) the glistening walls, the formations, and the immenseness of the entrance room. The others followed close behind, and more than a few oohs and aahs were heard.

At that time of the year, the interior of the cave was about 20° cooler than the outside temperature, and the relative humidity was very high. A few seconds after entering the cave, we were standing in a room which was possibly 100 feet across and 30 feet high. The floor was covered with sticky mud which clung to boots and smeared any clothing which came in contact with it.

We were still walking downward on a rather steep incline. The room we had entered was roughly circular, and we were walking through the dead center of it. At this point we had not entered any passageways (which I define as connecting tunnels between rooms). After only a few minutes we reached the opposite side of the room and looked back at the entrance. It was at this point that we realized that we were actually cave explorers. We were not so confident, however, as to stray more than a few feet from each other.

After looking the situation over for about ten minutes, pounding hearts eased up a bit and the old exploring urge shifted into full gear. Looking to the right, one of us spied a passageway about three feet wide and four or five feet in height. We worked our way over toward this opening from the entrance room and slowly entered after checking the ground for any signs of carniverous animal tracks. About ten feet into the passageway (whose floor was very dry compared with the main room), the ceiling began to lower and the walls began to close in. A few feet further in, we were down on our hands and knees and finally the tunnel came to an abrupt end. Striking out here, we made our way back to the main room, chose another passageway, and encountered the same results.

In all, we must have entered five or six tunnels and passageways which led off the main room. All ended shortly in solid limestone walls. Check as we might, we could find no other passageways off of this main room. Our cave exploring adventure ended exactly 23 minutes after it started. All the cave contained was this single room, a few dead-end passages, and a few stalactites. We were very disappointed, to say the least. Our full day of cave exploring had taken less than an hour, and the farthest we got was about 100 feet from the entrance.

This experience may seem unusual but it's not, really; many caves are little more than relatively shallow depressions in the limestone rock. Obviously, millions of years ago, the conditions of the limestone and the environment were not such that a large cave could form in this particular location. It must be remembered that the vast network of rooms and passageways which are Skyline Caverns lie only a few miles from this spot. We checked for any signs of cave-ins or mud-filled passageways, but could find none. Many years later, I learned that the rumors about Sink-Hole Cave being thought of for commercialization at one time were false and that the cave had never been more than a fair-sized hole in the ground.

In order to make the best of a bad situation, we did stay in the cave a few more hours, ate lunch, and practiced our rock climbing abilities on a jagged wall at one end of the entrance room. Looking back on the situation now, it could be said that this first experience, while disappointing, was probably ideal for three young explorers. This certainly was not a difficult cave by any stretch of the imagination, and we were allowed to become acclimated to cave exploring pursuits in an atmosphere of calmness (after the first ten minutes).

We knew full well that there were other caves within a short distance of this one. However, the rules of safety were adhered to even then. We had let other people know where we were, and it would be impractical to drive back to town to inform them we were going elsewhere. Also, we had not obtained permission to enter the properties where some of the other caves were located, so we decided to make the best of what we had. The picnic lunch underground even in this small cave was quite enjoyable and served as an indication of the fun we would have in a larger cave at some future date.

Working with local police, I had the occasion to enter this same cave about 25 years later in search of a possible murder victim. While nothing was found, being in that large room again brought back memories of that first cave exploring expedition. Nothing had changed much, except that a great deal of trash and debris had been thrown on the floor by thoughtless individuals. This is one of the unfortunate aspects of a cave entrance located so close to a highway and offers relatively easy access without the need for ropes and specialized equipment.

DEVIL'S CAVE

The following weekend, we three spelunkers who had conquered Sink-Hole Cave made arrangements with a property owner to explore a cave with a fairly small ground-level entrance. It was called Devil's Cave by local residents, several of whom had been inside. The real origin of the name never surfaced, but I would suggest that it might have come from the fact that the entrance was very tiny and required the spelunker to lay flat and wiggle through. It was probably less than a foot in height and maybe two feet wide. Even smaller individuals who entered or exited had a *devil* of a time.

As is always the case when properly planning an exploration, I checked with many people who lived close to the cave entrance in order to find out what to expect. The few who had entered indicated that we might find some bats, along with the standard formations, and that the cave wasn't tremendously long and could be fully explored in an hour or so. I also learned that most of the passageways were fairly tight and the largest room measured only about ten feet long. I was also told that the passageways were low and there was no point where a six-foot man could stand fully upright.

We entered the cave at 10:00 on a Saturday morning. The fear which sometimes comes over you when entering any cave is magnified several times when it becomes necessary to crawl through a tight opening at the base of a limestone ledge. While you are wiggling to get through, you are completely helpless should you decide to exit rapidly. Therefore, a focused-beam flashlight was directed through the entrance while I got down on hands and knees to see exactly what I would be getting into. I could see no signs of snakes or other "creepie-crawlies," but a long stick was inserted through the opening and stirred around on the sides just to see if we could elicit any movement.

That done, the moment of truth had arrived, and I laid flat on my back and inched through the opening feet first; I could not wear my hard hat while executing this maneuver. Once inside, my partners handed it through, whereupon I immediately activated the carbide lamp. I was in a narrow passageway about four feet wide and of about the same height. I could see some small stalactite formations and even a miniature column about five feet away. My partners hustled through the opening with a bit more bravery than I had displayed (but of course, they had a man on the inside by this time). In single file, we began to inch along the passageway in a full walking squat. This is often known as the "duck walk" technique and is extremely tiring but saves the knees from assorted bumps and bruises.

The rest of the cave was little different from the opening passage. The cave seemed to be one long winding tunnel which eventually led to a dead end. It appeared that at one time a passage led further, but a larger boulder was filling it now.

Backtracking, we paid particular attention to the ceiling, which was just a few inches over our heads while doing the duck walk. I had never seen a bat suspended from a cave wall or ceiling before and nearly missed the only one this cave contained at the time. It reminded me more of a stubby, wooly worm than it did a flying rodent, but there it was in the glow of the carbide lamp, hanging from the cave ceiling. It didn't seem to notice us at first, but suddenly, one eye opened menacingly and then the other, and it began to twitch a bit as it became nervous. We decided to leave well enough alone, as the prospect of contending with a scared bat in such cramped quarters was not all that pleasant.

We exited the cave about an hour after entering and proceeded immediately to another a few hundred yards away on the same property. This small cave had no name and we knew nothing about it except that some of the children in the area used it as a fort. It was very disappointing—just a single passageway which led approximately ten feet into the side of the hill. However, at the end of the passage was a tiny calcite formation that had been partially broken by the small cowboys and Indians who frequented this passageway after school.

ALLEN'S CAVE

Several other small caves were explored after our entrance into Devil's Cave, and after several months of this, we three intrepid spelunkers decided to tackle a larger project. Skyline Caverns owned a fair amount of property, and on one edge was another small cavern system known as Allen's Cave. This had been explored many times by cave clubs from the Washington, D.C. area, and it had even been written about in a few club publications. It was an easy matter to obtain permission from the corporation which owned this property to explore the cave, although it was necessary for our parents to sign release forms in case of an accident. This is pretty much a standard procedure when dealing with corporate properties.

It had been rumored that Allen's Cave did or at one time had connected to the passageways in Skyline Caverns. However, we were told by the corporation representative that this was probably not true, and no passageways had been discovered that would link the two together. They also informed us that extensive exploration had turned up no anthodite formations, although one room of this cave is situated probably less than thirty feet from an anthodite room in Skyline Caverns.

We learned that while it is possible to fully explore Allen's Cave without ropes, short lengths were recommended to make the going easier. Allen's Cave was laid out at several levels, but no drop was deeper than ten feet, and these could be circumvented if desired. Naturally, we were looking forward to scaling a few sheer ten-foot "cliffs" and felt would be good experience for more difficult exploration.

Even though the corporation had informed us that there was no natural link between the two caves, we had also heard rumors from the guides of Skyline Caverns that there was, in fact, a natural passageway at one time, but that it had been sealed shut with steel and concrete. We were determined to discover what the real situation was.

We wanted to allow ourselves at least three hours for exploration, with a contemplated maximum stay of five hours. This gave us the option of extending our explorations if desired. We were especially interested in seeing the largest room in Allen's Cave, which was called the Ball Room and supposed to be massive in size. This was also to be part of a high school science fair project, so knapsacks were packed with a number of plastic containers to allow us to bring back clay and water samples, later to be analyzed for any signs of plant and animal life.

At 9:00 on Saturday morning, we had hiked from the Skyline Caverns parking lot across a small mountain to the entrance of Allen's Cave, which lay on the side of a hill. The entrance was rather large, as we could walk in upright. This brought back memories of our first exploration in Sink-Hole Cave, where the entrance was very large but the actual cave was little more than nothing. However, once inside, we saw a long passageway whose ceiling extended up in excess of twenty feet.

At last we were in what could truly be classified as a "real" cave. After about ten minutes of travel down this passageway we came to a massive flowstone formation a full 20 feet in height, extending from the ceiling to the floor. Throughout the passageway, we encountered moderately large stalagmite formations with matching stalactites overhead. This was the kind of cave exploring we had imagined, and the type seen in all kinds of movies from *Tom Sawyer* to *Journey to the Center of the Earth.*

Shortly thereafter we rounded a corner and the ceiling got even higher, extending to about 30 feet. We felt rather dwarfed by the immenseness around us, and it was there that we discovered the missing link. In the corner of the jagged wall about ten feet from the floor was what appeared to be a horizontal passageway which seemed to lead toward Skyline Caverns. A foot inside there was a mysterious white substance with a rough exterior which appeared to be concrete. We made a note to ourselves to see if we could get one of our party up to that point on the wall for a closer examination on our return trip.

For the present, we had our first ten foot drop to contend with. The drop was more of a jagged cut horizontal to the passageway we were in that led down to a stream bed. The two walls of this cut were no more than four feet apart, so the rope was not necessary, as it was an easy "chimney walk" to the bottom. One by one, we braced our feet against the far wall and our backs against the near one and simply walked down the fissure to the creek bed. As one person would come down, the one below would stand clear in case any rocks or loose debris might be inadvertently knocked down. While a hard hat will protect you from some of this, there's almost no excuse from ever putting yourself in a position where you would be standing under something that had the potential of coming down on you. (In 20 years of cave exploring, I have never been struck on the hat by a loose rock. However, it has saved many a skull fracture when raising up too soon beneath a low-hanging rock ledge.)

Now in the fissure, we had a choice of two directions to take, left or right. We elected to go to the right and followed this narrow passageway. After about 25 feet the walls began to close in and further travel became impossible. Shining a flashlight beam down the narrowest portion showed that this passage seemed to end a few feet further in and that this was not some miniscule passageway into a large room.

An about-face was executed, and we started down the passageway which led more or less in the direction of Skyline Caverns. The fissure opened up into a moderate-sized passageway and began to rise slowly. We encountered a few jutting ledges which we climbed over with little difficulty. The floor was becoming very uneven at this time, being laid out at a 30° angle. Rounding a sharp bend, the passageway suddenly ended in a solid limestone wall about ten feet high. We knew *this* couldn't be all there was to Allen's Cave, so we backtracked 50 feet or so and spotted an overhead passageway perpendicular to the one we were presently in. The opening was located some five feet up the wall, which was not limestone but wet clay. Foot and hand holds had been dug beneath the opening, so it was an easy job to scale the wall. These had obviously been left by previous explorers.

The opening was about four feet wide and three feet high, so it was necessary to crawl through for about 15 feet. Gradually the ceiling rose and we were again in a fairly large passageway which twisted and turned deeper into the mountain. Since entering the fissure we had seen few formations, but in this overhead passageway, stalactites and stalagmites again appeared, along with a column or two and a fair amount of flowstone. We were constantly staring at the ceiling in hopes of discovering an anthodite formation. While we were told that none existed in Allen's Cave, we assumed that the Skyline Caverns corporation, which advertises their cave as the home of the world's only known anthodites, might tend to keep other discoveries a secret. (Of course, if this had been the case, we would never have been able to obtain permission to explore Allen's Cave in the first place.) Our teenage minds still had wild thoughts of making a fantastic discovery hitherto unknown to mankind. The closest we came to finding an anthodite formation was the discovery of some white calcite formations known as aragonite. These take many shapes but in no way resemble anthodites, except for their white color.

By this time we had been in Allen's Cave approximately one hour and fifteen minutes, and upon entering an enlarged area in the passageway, it was unanimously voted to take a break. We sat on our knapsacks and enjoyed steaming cups of coffee from a shatterproof Thermos. This brew along with a few chocolate candy bars warmed us and gave us the strength to start back in fifteen minutes later with renewed vigor.

Another fifteen minute walk brought us out into a large room which must have been the Ball

Room. We were somewhat disappointed because the room really wasn't gigantic in size. It measured about 40 in length, 25 in width, with a ceiling height ranging from 10 to about 20 feet. It was nearly devoid of formations and was simply just not that impressive. Of course, we had visions of a room the size of a supermarket parking lot, replete with at least 50,000 delicate stalactite formations and an equal assortment of stalagmites.

This would have been the ideal spot to have taken our coffee break, but since we had been rejuvenated only fifteen minutes before, we decided to immediately explore every nook and cranny in this room. In some ways it was reminiscent of the single room in Sink-Hole Cave, in that several promising-looking passageways could be seen leading off in several directions. Like Sink-Hole Cave, all of them went only a few yards and then abruptly ended. A small stream flowed near one end of the room, so we followed it for a ways until it ended in a small pool of water in a shallow depression. To be accurate, the stream was more like a fine trickle and probably vented itself through several tiny passages beneath the pool.

Cameras were brought out and several photographs were taken of members of our party, mud-smeared and jubilant. Another hour was spent looking for other possible passageways, formations, cave life, etc. Samples were also taken from the stream and surrounding mud for the science fair exhibit.

After this it was decided to head back toward the entrance in order to check out the "missing link" to Skyline Caverns and eat lunch at the mouth of the cave. The return trip was uneventful, although one's mind begins to play tricks in situations such as these. Sometimes we had the distinct feeling that we hadn't passed a particular point on the way in or that we had gone by a certain formation three or four times in the previous ten minutes. While we knew getting lost was practically impossible because only one fissure/passageway led in, the thought must have occurred to all of us at least once. When doubts were getting severe, we reached the opening (overhead) into the fissure, scaled the wall, and were once again headed down the stream bed. After about fifteen or twenty minutes we again encountered the ten foot drop, although now it was a ten foot cliff to scale. We went up the same way we came down, by chimney walking. It was a little more difficult going up than coming down, but no real problems were encountered.

By now we were thinking about the concrete-filled passageway we had discovered on our way in and just how we would go about getting a man up to the opening. The walls were far too wide to chimney walk and were not covered with a foot or more of clay as had been the case with the overhead passageway and the fissure which led to the Ball Room. We nearly missed the spot where the overhead passageway was located and it was necessary to backtrack about 20 feet to locate it. Another glance told us that we would have no problem getting up to it because a large stalagmite formation was situated on a bulbous outcropping from the wall just beneath the opening. This may sound a bit unusual, since stalagmites usually form on the cave floor. But remember, cave walls are not perfectly perpendicular to either the floor or the ceiling. They are complex masses of all sorts of curves and depressions. At this point, the wall jutted out and formed a fairly flat area for a stalagmite to form vertical to the floor, fed by the drippings from an overhead stalactite.

A section of rope was fitted with an appropriate knot to convert it to a western-style lariat. It took only a few tries to get it looped over the stalagmite, which was a full foot wide at the base. Strenuous tugging on the opposite end of the rope indicated that our wall-mounted anchor was adequate for the task of supporting one of us, and the smallest of our group began to scale the wall. It was at this time that one of the few accidents I have ever encountered in a cave occurred. No, it didn't happen to the man who was scaling the wall; I was the victim. In bracing the rope from the bottom, I slipped on a wet rock. Landing with my right leg under me, I knew for sure I had broken a bone and had banged my back pretty well also. Our man on the wall immediately descended, and he and the other spelunker pried my leg out from under me (slowly) and gave me an underground physical. Fortunately, it was quickly determined that nothing was broken, but I had suffered a bad bruise to my back and right knee when these portions of my anatomy came in contact with limestone (hard as rock, you know). After a few minutes of limbering up I determined that I was going to live and found that I could walk alright, although I exhibited a slight limp for a few days after the expedition. One of my caving partners suggested that it might be safer to let *me* climb the wall because there it would be impossible for me to trip over my own feet, but the original climber tackled the wall again while I paid more attention to my footing. He was up to the opening in no time and disappointingly discovered that what appeared to be a passageway filled with concrete was simply a depression in the wall, the back of which was composed of a very light-colored limestone. We had been told the truth. There was no natural entrance (plugged or otherwise) between Allen's Cave and Skyline Caverns.

Two of us walked and one of us limped back to the entrance of the cave. We had been underground a full three hours and it was quite nice to see blue skies again. By now the sun was high in the sky and the heat of the summer day felt good at first but then drove us back partway into the cave entrance, where the cooling 55° breeze was our own natural air conditioner. To date, this was the most exciting and complex cave we had explored.

VARIOUS AND ASSORTED CAVES

By the end of the summer we felt comfortable and confident in most aspects of cave exploring, so we decided to go on an overnighter, camping out and exploring as many caves as possible. We had chosen a site in the mountains where we had spotted a great number of caves. Since their entrances were located on several different properties, it was necessary to obtain permission from all landowners involved.

This expedition started on a Friday afternoon and would extend through the next evening. We pitched camp at about 6:30 Friday evening, which left us with ample daylight to explore the surrounding area. My hometown is located in a very prolific area for caves and caverns. To readers from less densely populated areas, it will sound almost unbelievable that from our campsite (in the middle of the woods), we could see the openings of more than a dozen caves. We did some exploring that afternoon, entering the caves which had significant openings that did not require strenuous activity to explore. We did not want to tire ourselves out for the following day, which was to be devoted entirely to spelunking. Our main purpose at this time was to decide which caves we wanted to enter on this outing. The fact is that there were so many caves in this area that it would be impossible to explore them all in a week, much less in the day we had allotted ourselves.

How do you decide which cave is worth exploring and which isn't? This is a hard question to answer; to some extent, experience pays off. After you've been in a number of caves, you can sometimes look through the entrance or go ten or fifteen feet inside and make a pretty fair guess as to whether or not the rest of it is worth taking a look at. This doesn't work in every case, but I have used this method many times with pretty good results. Another method involves tossing a rock into the opening and listening for sounds. This is especially applicable to caves whose entrances start out as a steep and steady incline, or a near vertical shaft into the ground. On this Friday evening we discovered an opening about two feet in diameter which seemed to run almost straight down. Tossing a heavy stone in revealed that this natural shaft went down for a considerable distance and stopped in a pool of water. We marked this one for definite exploration possibilities on the following day. By sundown we had mapped out ten exploring possibilities. The first one would be tackled near sunrise.

I should point out that this expedition planned to explore most of the caves during the day simply because that was the sleeping/working schedule we were accustomed to. Exploring a cave at night is identical to a daytime exploration after you get through the entrance; for individuals who sleep during the day and work at night, this is an ideal hobby, although finding those cave entrances at night can be a little difficult at mountainous sites.

This was a special expedition, because we three eighteen years old (by now) were tackling a massive assignment and even hoped to possibly find a cave that had not been previously explored by the many cave buffs who habitually frequented this area during the milder months. The camping experience was educational too, in that it prepared us for a planned overnighter inside a large cave sometime in the future. The evening was filled with cooking chores and preparing the equipment for the following day. Ropes were meticulously uncoiled and checked repeatedly for any sign of wear, then coiled again into compact loops for the actual exploring. Carbide lamps were gone over once again, the nozzles cleaned and the mechanism checked. By this time we had accumulated a fair amount of spelunking gear, including rubber knee pads, fairly clean coveralls, and even specially rigged photographic equipment. The latter was enclosed in a padded surplus ammunition box which originally held a .50-caliber machine gun cartridges.

While we intended to get to sleep at a decent hour, midnight saw us still lolling by the campfire, exchanging tales of what had been and what might occur tomorrow. We finally turned in and slept comfortably on inflated air mattresses.

We arose shortly after the crack of dawn, prepared a hasty breakfast, and then set out to explore what we decided would be the most difficult cave first. This is the one that consisted of a near-vertical shaft in the ground. Fortunately, due to our campsite selection, it was not necessary to haul a lot of gear a great distance. All of the caves were within a short distance of the campsite, so this served as our central supply terminal.

Hell Hole Cave (as it was later to be named) consisted of a vertical drop of about 15 feet, which then turned into a steep incline for another 25 feet. A well-placed tree next to the vertical entrance served as our anchor point for the descent rope, and I entered first aided by a flashlight tied to my belt which directed a focused beam straight down. At the bottom of the 15 foot drop there was standing room and I could see down to a pool of water (about six inches deep). I had actually entered through the top of a moderate-sized room which apparently had no horizontal entrances. Since the other two members of the expedition could not enter until I had reached the very bottom of the incline, I called to one of them to drop me my rubber boots. This accomplished, I dropped down into the pool of water. The room was very narrow, with most of its sides being taken up in the vertical plane. Once in the pool, I noticed a solid wall to my right and what appeared to be a moderate-sized passageway to the left. I also noticed the initials of other explorers which were smudged on the wall by means of a carbide lamp. (This is common practice with some spelunkers, and while it does not physically abuse the limestone rock, it does detract from the natural beauty, so this practice is not recommended.) Knowing that the cave had already been entered took some of the adventure out of the entire maneuver, but I continued to check out this single passageway.

Rounding a corner only a few feet from where I had originally splashed down, I found another solid wall. This was all there was. The cave which we had considered to be most promising was little more than a hole in the ground with a shallow pool of water at the bottom. I signalled my anchored partners on the surface that there was nothing really to be seen and began the trek upward. It was at this time that Hell Hole Cave was given its name. The cave was extremely wet, and in the short time I'd been down, the rope had become covered with water droplets. While the bottom of the cave was narrow, it was not so much so that the chimney walking technique could be used, so I was forced to come up the rope hand-over-hand. While the first 25 feet was more of a steep incline than a sheer drop, I was quite exhausted when I reached the top ledge. From this point on I could execute a partial chimney walk routine to get to the cave entrance. I say *partial* chimney walk because the vertical shaft was too narrow for me to actually brace my back and feet against the opposing walls. In this case, I used my knees against one wall and my back against the other and wiggled my way out with the aid of my partners who hoisted up the rope at my command. I was probably in this cave no more than 20 minutes, but it seemed like three hours by the time I exited.

I took a short breather while my two friends hauled gear to the next cave entrance. This one was a bit easier. The entrance was in the side of a hill, of comfortable size, and went down at a slight incline. We could enter standing upright, but gradually the passageway narrowed and split off into three smaller tunnels, one to the left, one to the right and the third in the center. The center passageway was the most interesting because it consisted of a narrow slit in a solid flowstone wall. Examining it with a flashlight, I saw that about three feet in, the passageway took a 90° turn to the left. Peering down that narrow passage I could see more flowstone. However, it would be very difficult to get through the slit and make the turn . . . and if the passageway narrowed at a further point it would be impossible to turn around. This meant that I might get in all right, walk ten or fifteen feet down the passageway, and then have to back out again. I was not certain that I could extricate myself even if I were able to get in. For this reason, this passage was never explored. Certainly it would have been next to impossible to have gotten ***permanently*** wedged in there, but the aspects of the situation made me uncomfortable enough to scratch the whole idea.

The passageway to the left was tried next. It became very narrow, but we continued moving for about 30 feet. At this point the ceiling got so low that it became necessary to lay on our backs and slide along on the cave floor with the ceiling only a few inches from our noses and the brims of our hard hats. The passage widened horizontally, however, and a number of stalactite/stalagmite formations were seen. We continued in this mode for perhaps a half hour and eventually the ceiling closed in to the point where we could go no further. Some difficulties arose here when it became necessary to do a 180° body reverse and wiggle out again. This took a few minutes, but eventually our heads were all pointed toward the way out. It took less time for us to get out than it did to get in, because we were more concerned with returning to the larger passageway than we were with exploring. This was a low spot in the cave, packed with mud and very moist. The slippery mud made the going easier, however, although we were completely covered by the time we returned to the large passageway.

At this point we took a ten-minute coffee break and then proceeded to explore the right-hand passageway. This one also narrowed but not nearly as much as the previous one, and it was possible to duck walk the entire way. The passage angled downward, and we eventually dropped into an office-sized room with a fast-moving stream at the bottom. The stream entered the room from a small fissure to the left and exited through a small passageway whose ceiling came to within six inches of the water. From the sounds of it, the stream was flowing very rapidly into a larger room somewhere down the passage.

The cave exploring books we had read told of spelunkers laying on their backs in ice-cold water and pushing themselves along with their faces kept barely above the surface and flush against a low ceiling. We gave some thought to doing that in this case, but we had many other caves to explore that day—and anyway, we were all afraid to try it. This was certainly a good decision to make, since it would have been impossible to see where we were going or even to light our way to the proximity of the ceiling to the water. This is an exploring situation for highly experienced teams who can respond quickly and smoothly should a problem develop. I had never heard (and still haven't) of any spelunker drowning in a cave, but I sure didn't want to be the trend-setter.

At this point we slowly exited the cave, looking for signs of overhead passageways. None were found and we saw blue skies again approximately two and a half hours after entering. By this time we were all a little tired, and since it was approaching the noon hour, we traveled the short distance back to camp and cooked a hot lunch. We had visions of exploring three or four caves before noon and were decidely behind schedule, although we would have been delighted to explore only a single cave if it had enough passageways to keep us busy all day long.

Lunch completed, we tackled our third cave —or maybe I should say we *started* to. The opening was about six feet in diameter and was located in a small depression that went straight back into the mountainside. We could see it about 200 yards away and just below our campsite. As we approached, the smell of decay became quite overpowering and it was obvious that there was something long-dead lying within. Just then the most gargantuan winged monster I have ever seen in my life exited the cave with all the dignity and aplomb of a wounded water buffalo. I cannot recall a time before or since when I have been more frightened. Needless to say, we never entered *that* cave and when we were able to swallow our hearts again, we identified the creature; forever after, that particular hole in the ground became known as Buzzard Cave.

Incidentally, occasionally a spelunker will run across some unusual animals in caves. Of course, bats and camel crickets are normal, but in certain situations, buzzards and other birds of prey will use them as dens. In one very remote area, I also discovered a fox den containing a single fox pup. The mother was either out hunting or had heard me coming and left the scene. Situations like this are fairly rare, but be on the lookout for other than expected cave fauna, as some unpleasant situations can result. The area where we were exploring did not contain any truly dangerous animals other than a possible black bear or two. Regarding the latter, the possibility of running into one in a cave is extremely remote, but we would always check the entrance of those few caves which offered openings large enough for a bear to enter for signs of tracks or droppings. Remember, however, that even a small animal such as a ground hog or raccoon can become rather nasty in an enclosed space with its back against the wall. This is something to be aware of but not to be overly concerned about. The fox cub and the buzzard were the only two animals I have ever encountered in wild caves (other than bats, camel crickets and the usual cave fauna) in my twenty years of active spelunking.

Since Buzzard Cave had been unanimously scratched (pecked) from the list of things to explore, we moved to another opening about fifty feet away. This one was small (about three feet in diameter) and entered at the base of a hill angling slightly upward. Admittedly, we were all a bit paranoid because of the experience at Buzzard Cave, so a number of rocks were tossed in the opening while we listened for any animal responses from within. When nothing was heard, the cave was entered, we knew we had found a good one. The three-foot entrance passage extended only six feet and then suddenly dropped about five feet into a large room. The ceiling was about 20 feet in height and the oblong room was about 40 feet in length. The floor was fairly level; at one end a large passageway led back into the mountain. In wet weather a small stream flowed through this passageway, but only damp clay was present at this time of year. We took some photographs of the room and then proceeded down the passageway, whose ceiling was about ten feet off the floor. It traveled a fairly straight path and closely resembled a picture I had seen in a book on cave exploring of spelunkers exploring Carlsbad Caverns in the early days of its development. This was exciting and several photographs were shot at this point as well. Several sub-passageways broke away from this main channel. The first one was checked out by one member of the party who found that it simply paralleled the main passage for about 15 feet and then entered again. The entire main passageway must have gone on for about 150 feet and all side passageways continued in the same vein as the first one.

After we had trekked the 150 feet it was necessary to climb over a large clay mound and then travel in a smaller passageway slightly above the first one. Here we found a great many delicate stalactites and stalagmites. Unfortunately, however, the passageway continued to narrow and eventually we could go no further. Crawling back to the main passageway, we made our way to the large room and finally exited again. This was to be the most impressive cave we entered that day. However, the giant proportions of the entrance room led us to expect far more than was actually there. While this cave only contained one large room, its large main passageway was even more interesting, and I was to return here many times in later years.

The remainder of the afternoon was spent exploring three more caves. One was quite small, but consisted of a number of narrow, snaking passageways which interconnected at a distant point at the rear. There was no point at which a six-foot man could stand upright, and we quickly tired of crawling around on our stomachs. Another cave consisted of a very large entrance, off which was a medium-sized room and several wide but shallow passageways. Still another seemed to consist of one long passageway and little else. A few small cubicles barely large enough to hold a grown man could be seen at eye level in the wall, but little else was found except a lot of white calcite on the walls. One unique sight was seen in this last cave. We were on our way out and I spotted a cup-like depression in the side of the wall. This may have been where a stalagmite had broken off hundreds of thousands of years ago, and a small basin about four inches in diameter remained. Moisture from the walls had collected here to form a tiny pool about two inches deep. The collected moisture had flowed down a calcite wall and tiny crystals were floating on the surface of the pool. These looked almost like thin ice on a pond after a partial freeze. The crystals did not form a solid layer, and when I put my fingers in the water I could feel their grainy texture. I have never seen anything like this since that time.

By now it was time to break camp, load up the jeep, and drive the five miles back to town. We lay in the sun for a few minutes and talked about the day's exploring. This act served two purposes: First, it gave three weary spelunkers a chance to rest and recuperate, and second it allowed the wet mud on our coveralls to dry so that most of it could be flaked off instead of smearing. The expedition was officially over; while we had not discovered an unexplored cave, we had seen many sights that we'd only read about and never before experienced, so we felt completely satisfied. This is what the expedition was all about, so it was termed a major success.

RODGER'S CAVE

In late August, all three members of our spelunking group were making preparations to pursue various fields in college, and our last planned expedition of the year was to be an overnighter in a local cave of much renown. It had been named Rodger's Cave after the previous landowner and was sealed to access by a steel grate which covered one of two original entrances. The other entrance had been sealed shut, apparently by blasting. This was a potentially dangerous cave because the only way in was by a 75-foot drop almost straight down. The 75-foot drop was actually in a series of three successive vertical drops of 25 feet each. The present owner would allow entrance only to those individuals who applied in person and who were willing to sign release papers. Many cave exploring clubs from all over the eastern United States had entered this cave, and rumors were circulated about a subterranean lake whose depth was never measured. It contained many winding passageways, one of which went for quite some distance into the mountain. This cave had the disinction of having been mapped and submitted to NSS (National Speleological Society). I had obtained a copy of the map prior to our first expedition into Sink-Hole Cave the previous spring. Additionally, the many passageways even had names. The long one was called Dead Bat Passageway, apparently named when the mapping expedition discovered one in this area.

Getting to the cave entrance was even dangerous, as it lay at the bottom of a large fissure in the earth surrounded on all sides by huge outcroppings of limestone. It was necessary to descend about 40 feet straight down to get to the cave opening. This initial descent was technically aboveground; imagine the entrance as lying at the bottom of a large limestone pit, such as might be found at a quarry.

I was known to the landowner, who had little more than a passing interest in cave exploring, but he was quite adept at judging an expedition by the equipment it carried. I approached him several weeks before we planned to enter and explained what we had done in the past and what we hoped to do in the cave during our overnight expedition. He immediately asked what equipment we planned to take and was quite impressed by our planning and forethought. He also made suggestions for several additional items we might need and pointed out a slight error on the map I had obtained. He was most concerned about our intended length of stay. A signed liability waiver would not provide absolute protection for him in case an accident occurred, a fact we both knew, and he was worried that if an accident occurred, we would not really be missed until sometime the following day. This problem was worked out by my promising to install a Rube Goldberg arrangement at the entrance which consisted of a 150-foot length of heavy twine which would be anchored to the top of the pit and extended down to the bottom of the 75-foot drop into the heart of the cave. At a prearranged time after we had entered, he would pull on one end of the string and we would pull back several times to indicate that all was okay. We estimated that it would take approximately five hours to fully explore the cave and set up a camp. Since we planned to enter at 9:00 on a Saturday morning, the signalling time was set for 3:00 sharp. He also promised that if we were all right but failed to arrange our schedule so that the signal could be given, he would personally keelhaul all three of us. With that, he persented me with the key to the grate which covered the entrance.

This was to be a very complex exploration due to the amount of equipment necessary to get in and out of the cave, the camping gear, food, and all the extras which had to be taken along. Naturally, each man could not be overloaded, as we would be walking a very long distance. Carbide lamps are fine for exploring, but for a subterranean campsite, lanterns and other battery-powered forms of illumination are mandatory. While the ventilation in Rodger's Cave was good, it would be very dangerous to even think about starting a campfire. We did, however, take along a Sterno cookstove which could be used for warming our supper.

While we were all eighteen and all parents concerned had become accustomed to our weekend outings, the idea of an underground overnighter was far more difficult to sell. The signalling system that was worked out with the landowner assuaged most fears, however, and the expedition was begrudgingly approved. This was mandatory, since at this time a person had to be 21 before he could legally sign a contract or a waiver of liability. Up to this time, only one other expedition had spent the night in Rodger's Cave, and this was composed of the adult members of a cave exploring club from Pennsylvania.

There was some thought given to taking along a pup tent for each individual, but this was scratched because of the massive amount of equipment which was absolutely necessary and the fact that we'd be covered by several feet of rock anyway. In place of these tents we took nylon sheets that would be supported above each individual's sleeping bag by small aluminum rods in order to keep off any water which might drip from the ceiling. Air mattresses were mandatory for a comfortable night's sleep; these could be stowed in a small physical area. One of the most important items we carried was a plastic trash bag in which to toss all of our trash. It would be necessary to refill the carbide lamps many times, and each fill requires that the expanded carbide ash be discarded. Too many spelunkers simply dump this in the cave, where it becomes a source of poisoning for bats and other cave life.

Planning the menu was a real problem because of our limited cooking facilities. It was absolutely mandatory that we be able to brew coffee and warm up some of our canned goods to fight off the chill brought on by extreme humidity and a year-round constant temperature of 54° Fahrenheit. All canned goods were removed from their original containers and stored in heavy plastic bags to make them more compact and transportable. Hot dogs were chosen as the main supper food because they could be easily cooked, and many could be stored in a small container.

Another problem encountered when planning an overnight stay completely underground is that of damp clothing. When exploring most caves, you traditionally get wet to the skin. This is all right for short periods of time but is highly uncomfortable for more than 24 hours. For this reason, each man carried a complete change of clothes, and our caving coveralls were treated with waterproofing compound and sealed at the cuffs and sleeves with rubber bands to prevent infiltration. The same was done to the sleeping bags, which were rolled very tightly to decrease their physical size.

For this expedition we carried the normal complement of rope plus a 250-foot length of heavy twine, 150 feet of which would be used at the entrance. We hoped to use the remaining 100 feet to judge the depth of the subterranean lake.

After a full week of gathering up, packing, unpacking, and repacking equipment, we were ready to go and arrived at the cave entrance at 8:30 a.m. The landowner was still asleep, so we left a note on his door (as promised) telling him that we were on schedule and expected to enter the cave at 9:00 sharp. We also repeated our instructions about the 3:00 signalling time. While we carried the normal amount of rope into the cave, an extra 150 feet was needed just to get into it. This would be left in place for the climb back. A 50-foot length of one-inch rope was tied around a sturdy tree at the top of the pit and dropped downward to the steel grate covering the opening. All knapsacks and equipment were lowered to that point with additional rope. This was hauled back in and recoiled for use in the cave.

Free of our burdens, we made the descent to the opening. Being the first down, I was securing the rope at the bottom to ease the passage of my partners when I received a scare which was reminiscent of the experience at Buzzard's Cave. Straddling the grate which covered the three-foot opening to the vertical shaft into Rodger's Cave, I looked down and saw a black spider which measured a full foot from leg to leg. It took my mind about a half a second to register that this monstrosity was a steel ornament attached to the grate, but unfortunately, my body reacted more quickly and I flung myself backward into a pile of leaves. For a minute it seemed as though disaster had struck even before we had entered the cave. My two partners descended the taut rope snickering.

The next problem came in getting the lock on the grate to open. It was a bit rusty and required a lot of twisting, turning and pulling. With the grate open, we were treated to the sight of the most vertical natural shaft I have ever seen. It seemed to go down forever, but with the aid of a flashlight we could see that it ended at a level floor about 15 feet from the opening. The equipment was rearranged and a 100-foot length of one-inch diameter rope was secured to a steel rod which had been previously driven into the concrete for this purpose. To make certain that the system was failsafe, the rope was also wrapped around a large rock, giving us two anchors for added safety. The rope was dropped down the shaft and I entered with little difficulty, going hand-over-hand to the bottom. The knapsacks were lowered and my partners followed shortly thereafter.

We were standing in a very narrow passageway whose ceiling was about 15 feet high but whose walls come to within a foot and a half of each other at various vertical points. We strapped on our knapsacks, struck our carbide lamps, and looked the situation over. Thinking back on this experience, I can recall the tremendous draft which came out of the depths of the cave and vented through the entrance. It was strong enough to cause the flames from the nozzles of our carbide lamps to flicker. Looking forward we could see a slight incline and then suddenly a drop. The remaining rope was tossed outward, and we could hear it fall all the way to the bottom of three successive drops. We inched toward the first one and with one futher tug on the rope to make certain it was secure, I eased over the edge. The first drop was not nearly as bad as I expected. It was purely vertical for the first ten feet but then began to slope out into a steep incline. This went down for another 15 feet and then there was another vertical drop. This one extended a full 25 feet and ended in a narrow ledge. At this point I could stand upright and take a short breather. I noticed that the walls were ideally separated for our chimney walk routine and saw that it would be possible to use this technique all the way down and back up again if legs and back held out. This could be used in an emergency but would certainly not be recommended for normal practice.

It was at this point that I heard the audible squeak of bats as they were gearing up to go ultrasonic. I looked below my feet and saw five or six bats flying erratically in the space which encompassed the final 25-foot drop. While bats are nothing to be frightened of, they can become a nuisance in enclosed areas and will fly into you, regardless of what the experts say about their "radar." I took a deep breath and began to descend the last 25 feet, which was more of a very steep incline than a vertical drop. When I was halfway down a bat flew into my right leg. I glanced downward and then looked up again to see one flying straight at my face. All I can remember is seeing those tiny fangs in the glow of my carbide lamp, gripping the rope with one hand and putting my other up to shield my face. Unfortunately, my hand was directed too high and momentarily came in contact with the flame of my carbide lamp. This caused a nasty circular burn and I could smell the distinct aroma of singed flesh. I quickly descended the remaining dozen feet, thankful to be down in one piece. While this may sound like a dangerous situation, it should be remembered that I was equipped with a safety line that was held by my partners.

The bottom of the drop was a little wider at the top and was covered with a great deal of rocks and other debris, making footing uneven. I let my partners know I was down safely and asked them to toss me the signalling twine which had been attached to the same tree that secured our original descent rope. You might wonder why we didn't simply use one 150-foot length of one-inch rope to go from the top of the pit to the bottom of the cave, and for signalling purposes as well. It looks good on paper, but the weight of this much rope is fairly great, and it would be very difficult to signal with it by pulling back and forth. The twine was tossed (in an outward fashion) and landed at the bottom without any serious entanglements. This would later be stretched a little further into the cave and then secured near an area where the map indicated we might find a good camping spot. The next man descended without incident and then the third. The last partner was the best climber of the group. He came last because there was no practical way to fit him with a safety rope. While I can only speak for myself, I feel reasonably certain that all of us looked back up that 75-foot drop, hoping and praying that we had enough expertise to climb back out again and that some kid didn't come along and untie our descent rope at the top of the pit. Fortunately, the cave was located in a well-isolated area and the landowner had assured us he would check in around our planned exit time in case any problems had developed aboveground.

Having descended into Rodger's Cave, the first order of business was to attend to my burn. It was a bit painful at first, but some ointment and a adhesive bandage from a well-stocked first aid kit took care of the situation. It was time to push onward, as a full half hour had elapsed since we had actually entered the cave. The base of the drop was a small stream bed which contained a small quantity of water. We moved forward following this passage which continued to be rather narrow for approximately 25 feet. The ceiling was high enough, so crawling wasn't required. Each of us had to walk slightly stooped over, however. Suddenly, the passageway widened and the stream ended in a long pool of water approximately one foot deep. There was enough space on the right side of the pool for a narrow path, and we paused for a few minutes to look for signs of aquatic life with a flashlight. Seeing nothing, we moved on and noticed the passageway ceiling getting lower. Ahead, we spotted what appeared to be a dead end, although a closer examination revealed a small opening in the upper right-hand wall. As we got closer to it we could hear our voices reverberating in what sounded like a moderate-sized room above. When we crawled through this opening, we felt like we were truly in a large cavern system. The room was not especially large (about the size of an office), but it contained a large opening at one end which revealed an even larger room which seemed to go on for quite some distance. This portion of the cave was drier than the portion we had already traveled, and we were treated to the sight of tremendous stalactites, stalagmites, and vertical columns. We saw very few flowstone formations, however.

The floor at this point was very uneven, and we were compelled to climb upward, walking over very large boulders which lay everywhere. Climbing out of the smaller room, we entered the large one and found rather frightening evidence of a single section of ceiling which must have weighed over 100 tons that had fallen thousands of years ago (we hoped). This large room was very cluttered and contained its own hills, valleys, and other features which we viewed from our vantage point. We were situated slightly above the room and it was necessary to walk downward in order to get to the floor.

It may sound like we were standing at great heights above this large room. However, we were no more than ten feet above the floor and worked our way downward until we were standing on fairly level terrain. In the underground, distances and sizes can be quite deceiving. All of the descriptions of room sizes and passageway lengths given in this chapter are my best "guesstimates" and could be way off. Depending on the angle, the light, and the surroundings, a stalagmite some ten or fifteen feet away may appear to be six feet or more in height but in reality is less than two feet high. Likewise, a large column at some far-off point may appear to be only a few feet high while it is really twelve feet or more in height.

What impressed me most about Rodger's Cave was the high ceilings. In the larger rooms, ceilings were 20 or more feet in height and only in a few areas was it necessary to stoop over or duck walk. We made our way through this room and decided that this would be an ideal place to make our camp for the night, owing to a fairly level, clear area in the middle of the floor. This would also give us the opportunity to dump a great deal of our gear and make further explorations with a little more ease.

We had some problems at this point. After setting up camp, we continued to explore and were specifically looking for a passageway that led off to the right of the large room which, in turn, led to the underground lake. As I said earlier, the room was like a world unto itself, having its own hills and valleys and even a small mountain range right down the middle which led up to near the ceiling. For about 30 minutes we kept getting lost in this large room. The map indicated a passageway somewhere to the right of the subterranean mountain, but we couldn't find it; explore as we might, we always ended up back at the campsite. Finally we discovered a fairly large boulder which we had not trod over before, and near it was a streambed (dry) which led down a passageway with ceilings approximately seven feet off the floor. In some places this passage was more than six feet wide. The going was rather rough because of the quantity of stones and stalactites in the path. It was also necessary to climb uphill in a few places and to negotiate a few four and five-foot drops. A few side passageways ran off the main artery but did not seem to lead to any other rooms . . . only dead ends.

After walking for about 20 minutes we suddenly rounded a corner and there was the lake. It could be seen through the aperture created by two walls that had suddenly come together about six feet off the floor. When the term *subterranean lake* is mentioned, you might think of some great expanse of underground water measuring several hundred feet across. This was not the case; underground *pond* would have been a more appropriate name. It was roughly circular, and I would guess about 25 or 30 feet in diameter. The lake was situated directly at the end of the passageway in a moderate-sized room with a 15-foot ceiling. Beyond the lake could be seen more passageways, but there seemed to be no way to make our way around it due to the steep limestone slopes which rose up from the water's edge. The map revealed that there was another passageway off the one we had just traveled, so we decided to try this one in order to circumvent the lake. Before doing so, however, we took quite a few pictures of the water, which appeared to be dark green under the directed beam of a flashlight and carbide lamp. Using a walking stick that one of my partners carried to measure the depth, we found that the lake must certainly be a deep pit which had filled with water. Unlike aboveground lakes and ponds, this subterranean body of water did not start out shallow and get deeper as you approached the center. It seemed to be bottomless just a few inches from its edge. The remaining twine was brought out, and a lead weight was tied to one end. Tossed in the middle, the weight dropped straight down but caused the twine to get hung up on a rock outcropping which lay above the water and directly in front of us. We were never able to get what I would call an accurate sounding. The line was quite tangled by now, and it was necessary to continually swing the weight out over the rock ledge, which would cause the twine to bind. If the weight had been larger, we might not have had this problem. Although this is just an estimate based upon many attempts, I would say the lake was at least 30 feet deep—but for that matter it could have been 230 feet. In any event, it was certainly the deepest body of subterranean water I had ever encountered up to that time.

By now we had been in the cave approximately two and a half hours, and we knew it would take us at least 20 minutes to return to the campsite. We hurried back up the passageway, continually looking for the side passageway indicated on the map that would take us across the lake. We especially wanted to get to this point, because our map had a big question mark on it in the passages indicated past this body of water. This meant that they had not been fully explored by the survey team.

We would have missed the passageway because it was located some six feet overhead. Fortunately there were hand and foot holds cut into the limestone wall by some earlier expedition. Up and over we went into a passageway which required a bit of duck walking until finally we could stand in a crouched position. When I say the passageway went over the lake, I mean this quite literally. Suddenly the passage ended in the lake room, and there was a four-foot jump to a rock ledge which continued into another passage. The jump had to be made over a finger of the lake which jutted out from the rest. We could all swim well, but the prospect of falling into that bottomless cistern was a little more than scary. Of course, a four-foot jump is nothing, really, so one by one, we got our nerve up and over we went. This also gave us something to think about (or maybe I should say worry about) on the return trip.

Having circumvented the lake, we encountered the wettest, muddiest portion of the entire system. This was due to several tiny streams which must have fed the lake proper. The streams were so small that they were only a very slow trickle of water emerging from the wet clay here and there. This was a very fascinating part of the cave, because it was honeycombed with many, many small passages which required a full prone crawl to explore. We crawled through every one of these and determined that all of them led to dead ends.

Each passageway we traveled narrowed to a point where we could go no further. The limestone in this area and the few formations we saw seemed to be much older than those in the rest of the cave. This may or may not have been the case, because while the floor was wet from the tiny underground springs, the walls and ceilings were geologically dead. This means that water had stopped dripping from them, and no new deposits of limestone were being built up. The few formations we saw were not shiny like those which are still growing due to the water that constantly flows over them. These were dull and appeared to be rough and chipped.

We spent only a short time here because of the shrinking passageways, then decided to return to camp for lunch and to take a breather before heading down Dead Bat Passageway, which lay somewhere to the left of the big room. The jump across the lake was not quite so frightening as the first time, but it was awkward to get out of the overhead passageway and down into the main one again without a rope because the hand holds were harder to find on the way down. On our return trip to the campsite, we noticed where some explorers had written their names on the walls with their carbide lamps and where some had even dumped large quantities of spent carbide on the cave floor. We stopped near one of these dumps and recharged our own carbide lamps (something we had done twice previously since entering the cave) and tried to scoop up this carbide from the earlier expedition and put it in the trash bag we had brought along especially for this purpose.

By the time we had reached camp, we had been in the cave approximately three and a half hours but were not feeling especially fatigued. We were hungry, however, since, it was slightly past noon. We decided to cook lunch at this point and save Dead Bat Passageway for later on. This would give us a chance to warm up, rest and relax, signal our man on the surface, and do some more exploring in the big room before heading out to a different part of the cave. We discussed the idea of skipping lunch and tackling Dead Bat Passageway immediately, but I had become paranoid about the the possibility of missing the prearranged signalling time with the landowner and decided it would be best to stay near the signalling cord. We had always made it a practice never to split up, since safety demands that, groups of three explore caves and always remain in close contact with each other.

While some spelunkers rave about the delicious meals they are able to prepare on a Sterno cookstove, I was not that good a chef, and it seemed to take a terribly long time to bring the water in our collapsible camp skillet to the boiling point so we could cook the hot dogs. After about 40 minutes, however, they seemed to be doing all right and I had even invented my own disposable hot dog bun steamer. This consisted of two aluminum foil pie pans, the bottom of one having been pierced several times with the tines of a fork. The other was used as is but was fitted with a crude hinge which attached to the edge of the perforated one. The rolls were placed in the bottom of the pierced one and the top one closed in on top to form a lid. When this entire mechanism was placed on top of our camp skillet, the steam from the boiling water rose into the area where the hot dog buns were and made them warm, moist, and tasty in just a few minutes.

As soon as the hot dogs were done, they were immediately eaten while the skillet was dumped and filled with pork and beans. When these were warm, they were also eaten and more water was added for another round of hot dogs. When cooking on a single burner, the rotation method is the only one which will work.

After our noontime meal was complete, a small coffeepot was put on to perk. This process would take quite some time, so we took turns exploring various portions of the big room. We were always in sight of each other, so we didn't really split up the team. A close examination of our campsite showed that a few bats had attached themselves to the ceiling, but they seemed to be calm for the moment and we hoped they would stay that way. We didn't find it necessary to use the nylon sheets as a canopy over the sleeping bags, because we had chosen an area where water was not dripping from the ceiling. The clay below us was quite dry. The only real problem was the dampness of the air. This tended to make the 54° temperature seem like 34°. Sweaters beneath our coveralls did a fair job, but exposed areas of skin required constant rubbing when not engaged in active exploring. The strenuous exercise often required in exploring a cave of this complexity can be a problem in itself for an overnighter. It is necessary to dress warmly (naturally). However, when you're exercising the warm clothing often causes excessive perspiration and you feel like you're freezing later when you stop to rest.

The only really wet area of the cave we had encountered so far was in the passageways past the lake. The waterproofing on our coveralls was doing an excellent job, so we were still fairly dry. Our coveralls were quite muddy, however, and these were removed at camp in order to avoid soiling our sleeping bags which we used as seats. It was decided to use these same coveralls for the later exploring, as it might be necessary to change into the dry spares for the trip out the next day. So far, we were dry enough and the spare change of clothes was not needed. (As it turned out, we never needed those extra clothes, because Rodger's Cave was unusually dry.)

By the time lunch was over and the campsite cleaned up, it was about 2:00 p.m., one hour away from signalling time. We continued to examine the different parts of the large room, looking for any unusual sights we had missed before. I was particularly interested in the small stream we had seen on the way in. Perhaps it would contain some blind cave fish. Since the signalling twine was anchored a few feet from the beginning of the stream, I suggested that we make our way there and examine the stream as closely as possible. We arrived with about a half hour to spare and concentrated the beam of an electric lantern into the water. Absolutely nothing could be seen, so I turned over small stones on the bottom in hopes of finding some kind of fauna beneath them. Suddenly a tiny crayfish darted out from under one of the larger stones. His reverse path was quite easy to follow in the crystal-clear water, and I was able to pick it up between my thumb and forefinger. I had seen pictures of blind crayfish found in a cave in Kentucky, and I was elated because I thought I had found the first one in Virginia. However, this was a normal, everyday, run-of-the-mill crayfish and had probably entered the cave from above by means of a tiny stream or passageway that filled with water when it rained. It had eyes, had not lost its pigmentation, and appeared normal in every way. It was placed back in the stream and further searching began, but nothing was found. Ten thousand years from now, the descendants of that crayfish may be blind albinos.

Right on schedule (3:00 p.m.), I felt a tug on the signalling twine and tugged back vigorously several times. With this job out of the way we were now free until 10:00 the next morning, when we were to check in with parents after exiting the cave. The landowner would call them with the information that we were safe so far.

We returned to the campsite and prepared to explore Dead Bat Passageway. It took us about 15 minutes to locate it, and in we went. It was very long and fairly straight, but nearly devoid of formations and certainly not the high point of the entire exploration. The ceiling was high enough to stand upright in some places, but it would suddenly lower and make it necessary to duck walk or crawl. We must have spent about an hour in this passageway, then exited and decided to return to the lake to collect some water samples and check for subterranean marine life. We decided to use the overhead passageway so that we could peer down into the small section of the lake that we had jumped over during the morning's exploration. We lay in the prone position at this small chasm and stared down into the water for what seemed like hours but must have been only 15 minutes or so. Bread crumbs were dropped in hopes of attracting any fish that might be present, but they never appeared. Since that time I have talked to others who have explored Rodger's Cave and no reports have ever been obtained of any marine life in the lake. As a matter of fact, no marine life was reported anywhere in the cave except the lone crayfish I had disturbed.

By this time we were past due for supper and getting a bit fatigued from the day's strenuous activities. We were a bit scratched and bruised from all the crawling and climbing, but no one was anything but jubilant. (Even the burn on my hand was giving me no problems.) We made our way back to the campsite and began to prepare more hot dogs and beans for supper. By this time I had become an expert at underground culinary preparation. Supper was even better than lunch, as it was topped off with large slices of chocolate cake which had been packed in a firm plastic container.

During supper we talked about the events of the day and checked on equipment that had been used. All of the flashlights and lanterns still had adequate charges. Carbide lamps were emptied and charged once again. We were running a little low on drinking water, so a quick trip was made to the small stream to get a supply of water for our carbide lamps. I'm sure this water would have been fine to drink, as it came from several natural springs, but we stuck to the canteen water we had brought along for drinking purposes.

Returning to the big room again, we stripped out of our coveralls and decided we could expend a few candles in order to get an idea of just how big the room really was. Unfortunately, in a large room, you can only see a small portion at any one time because of the limited light which is carried with you. What we did was place a candle against four different wall areas, with ourselves located near the center. We could then get an overall idea of relative room dimensions by looking at the candle lights. I would guess that the room was at least 60 feet in diameter. What made it really interesting, however, were its many features. In addition to the hills, valleys, and mountains already discussed, there were two massive columns from floor to ceiling near the center of the room which were about six feet apart. They seemed to form a natural gateway into the deeper portions of the cave. There were very few calcite formations, but a good number of stalactites and stalagmites of quite large size could be seen. All in all, Rodger's Cave had fewer formations per square foot than most of the other caves we had explored, but its massive size made up for this many times over.

By 9:00 that evening we were exhausted. It didn't seem as though we'd been underground for twelve hours, simply because lighting conditions were the same when we went in as they were at this time—and would be when we left the next morning. The dampness was really becoming bothersome, so we decided to turn in for the evening. The waterproof sleeping bags had been kept tightly zippered and were very dry on the inside. This helped to cut the chill and we slept very well in the absolute darkness which permeated everything when the lights were switched off. I was concerned with oversleeping and woke several times during the night to check my illuminated wristwatch. We were all up before 6:00 the following morning.

Breakfast consisted of hot coffee and cold donuts. This would be enough to hold us until we made our way to the outside. We purposely broke camp at this time and got everything packed and stowed so we could explore until 8:30, our planned departure time. This took about 20 minutes, so we had a good hour and a half until we would have to make our way toward the entrance. We had already explored almost every alcove in the large room and in many of the side passageways. This last period of exploration was done in a melancholy sort of way, resembling a last goodbye to an old friend. We made sure that we had not left anything behind and resisted the temptation to knock off a souvenir stalactite. Finally, it came time to strap on our knapsacks and make our way to the foot of the 75-foot drop.

Admittedly, the climb back to the surface was something I had thought about time and again during this exploration with a bit of trepidation. Thank goodness we had used the thick, one-inch rope rather than the half-inch manila that was far less expensive but extremely awkward for climbing long distances. On the way to the drop, we disconnected the signalling twine. It would be pulled up when we exited the cave.

It's hard to describe what went through our minds when staring up the drop which we had descended the day before. While claustrophobia had not set in, I believe we were all quite anxious to see blue skies again and get out of the infernal dampness. With a hope and a prayer, I grasped the rope and began to chimney walk up the first 25-foot drop. By the time I reached the top, I was quite happy because it had taken much less time to go up than it had to come down. In another few minutes I was at the top of the middle drop; five minutes after that I was staring up through the spider-protected grate at the blue morning sky. I could also see our entrance rope securely tied as we had left it and the other rope still hanging down into the pit.

I called to my partners, and one by one they climbed up and out. We did not use a safety rope on the journey up. Perhaps we should have, but we had learned by that time that it was not that difficult a climb and indeed, it would be possible to chimney walk the entire distance. Again, the trip back up was easier than the trip down, and this was always the case each time I explored Rodger's Cave in future years. When all the party was at the top of the series of drops, I went hand-over-hand up through the opening and stood in the sunlight which filtered down into the pit. It was a feeling of great relief and freedom in some ways, and also a feeling of sadness about leaving the most exciting cave I had ever been in. I had left my knapsack at the bottom of this 15-foot drop into the first level of the cave, so it and the other two were tied together and hoisted up into the pit before my partners exited. We encountered a slight problem at this point, in that the rope had apparently become snagged on a rock somewhere below the second level drop. It took about five minutes of tugging and twisting before it finally gave way and was pulled 75 feet to the surface. My friends exited the vertical shaft before the rope was pulled into the pit.

We were all quite fatigued from the previous day's explorations and from the journey to the top, so we rested for about ten minutes, gathered all our gear in a bundle, and pulled out the signalling string. Then we climbed to the top of the pit and hoisted our gear up after us. The jeep was located nearby, and all our gear was haphazardly tossed in the back. This expedition was completed, but we were to explore Rodger's Cave several more times as summer vacations and schedules permitted. Even today, this particular expedition was one of the most exciting ever.

OTHER EXPLORATIONS

Since those teenage years, I have explored many other wild caves and caverns in Virginia, West Virginia, and other eastern states. One cave had its entrance in the side of a sheer 100-foot cliff. To get to it, it was necessary to rappel down the face and then swing out and into the tiny entrance.

To make it even more difficult, the entrance sloped sharply upward (back toward the top of the cliff). Even worse, the cave only went back for 25 feet and then quit!

In another cave, I found the initials J.E. chipped into a limestone wall and beneath it the year 1749. To this day, I don't know whether someone in 1749 whose initials happened to be J.E. explored this cave and left his mark, or if someone in fairly modern times was playing a practical joke. The letters and numbers looked as though they had been there for some time, but it's difficult to tell because little aging takes place in most caves over the span of a few hundred years.

Once while exploring some undeveloped passageways in a commercial cavern, one of my partners and I discovered a prehistoric bear tooth—or at least that's what we *thought* it was. We had it checked and found that it was indeed a bear's tooth, but there was nothing prehistoric about it. It did, however, appear to be very old and could have dated back to the first English settlers or before. The bear was not actually in this cave in all likelihood. The tooth had probably been carried in by aboveground streams which flowed into this vast cavern network.

These stories of explorations could go on and on, as they sometimes do in the late winter months in front of a roaring fire with the faces of two or three hypnotized youngsters staring on. More often than not, a few hypnotized adult faces can be seen also. The fact is that I have had a great deal of fun exploring the underground, and it's given me a fair amount of satisfaction. I have never experienced any serious injuries in a cave, nor do I know anyone who has. It's worth repeating that the most dangerous part of a cave exploring expedition is the automobile trip to the entrance. Of course, like most pursuits, you can go about spelunking in the wrong way and wind up getting hurt. For example, it would not be appropriate for a beginning spelunker to tackle something like Rodger's Cave his first or second time out. Also, if you're unusually afraid of heights, you would want to stay out of situations which require any sheer descents. Most people do not have a fear of *heights*, but rather a fear of *falling*. A person who is afraid of heights is the one who can't look down at the earth below when in an airplane. We all have a fear of falling, but when we know we can't fall, the heights alone do not scare us.

Health enters the picture too. Those who are very overweight or do not have good stamina should stay away from this sport altogether. Cave exploring is strenuous work. You can get an idea of how strenuous it is by crawling under your bed and other household furniture for two or three hours at a clip. Try crawling up your basement stairs once or twice as well. This is a close simulation of what is encountered in exploring many wild caves.

Claustrophobia is another condition that can keep one from becoming a spelunker. Almost all of us suffer from this phobia at times, but it can be controlled by staying calm and continuing to reason. Claustrophobia can occur especially when one is mentally or physically fatigued. I have felt it several times toward the end of an expedition. I imagine it would get increasingly worse as you continued to remain underground. This type of claustrophobia is fairly normal, however, as I have also seen the other kind which manifests itself physically with nausea, disorientation, etc. Anyone who has it this bad already knows it and will experience it almost immediately upon entering a cave or other confined area.

Most people, however, young or old, can enjoy spelunking. Your age and physical condition will determine just how much you can do and the difficulty of the caves you wish to tackle. Whether you stick to the easy expeditions or try a really hard one, I guarantee you'll come away with enough stories to fill up not one but many winters by the fireside.

Chapter 6

Cave Courtesy and Safety

Safety has been stressed throughout this book along with courtesy when exploring wild caves. There is nothing particularly dangerous about cave exploring, although it's a difficult job to convince some people of this. A lion tamer was once asked about the safety aspects of his profession, to which he replied, "There is nothing particularly dangerous about putting your head in a lion's mouth, but the whole procedure is rather unforgiving of mistakes." To me, it is far more dangerous to stick your head in a lion's mouth than to put it through the opening of any wild cave in existence. As with any sport, profession or hobby, however, there are certain safety aspects that must be observed.

CAVE-INS

The biggest fear of non-spelunkers is the possibility of a cave-in. Natural caves are formed by the erosion, flaking, and even gouging out of soft layers of limestone material by flowing water. This process in itself usually eliminates the weak areas and leaves only the hard shell. The word "cave-in" is a misnomer; it should really be "mine-in." Mines are not natural caves. They are tunnels which have been cut into the earth by man. Mine tunnels may not be completely surrounded by hard rock, and their ceilings and walls often must be supported by a series of braces. These serve to keep the tunnel open. The natural process has been eliminated here as man cuts directly through to the ore vein. Weak sections of walls and ceiling are not removed by this process, as is the case with the natural formation of a cave. Add to this the fact that the ore is continually removed, further weakening the opening, and one can understand why mishaps occur. This is not to say that sections of ceiling *cannot* collapse in caves, but I have never experienced this nor even heard of it in modern times in wild caves.

During the rainy season, additional water may be flowing through caves that have natural streams and even rivers. At these times the erosion process is stepped up and there is a greater chance of an internal structural failure. Some caves whose passageways lie near the surface of the ground can be partially affected during freeze/thaw cycles, but these instances are quite rare. During my many years of cave exploring, I have never witnessed so much as a single small pebble falling naturally from the ceiling or wall of a cave. I have been pelted a time or two with small debris which was kicked loose by a spelunker above me; this is the reason for hard hats. After twenty years of exploring many of the same caves over and over again, not so much as a single, small-scale fall-in has ever been noticed. This speaks well of natural construction.

Several decades ago, the entrance of a cave was uncovered in one of the southern states when the land was being excavated for a building. When this cave was explored, the remains of an ancient Indian were found crushed beneath a very large boulder. Further investigation showed that the Indian had been digging into a clay bank which partially supported the boulder and it came down on him. Proper spelunking does not entail taking anything out of a cave that you didn't come in with. Therefore, you won't find this type of excavating going on while cave exploring.

Naturally, some caves have been purposely filled in by heavy machinery. This is often the case when building construction is going on in an area which contains a fair number of natural caverns. Some minor slides have occurred in commercial caverns during the process of commercialization where some passageways are widened, occasionally by blasting. These are all planned for, however, and rarely does anyone get so much as a scratch.

Another fear of some is that an overhead stalactite will fall, impaling them from end to end. It must be remembered, however, that stalactites (and stalagmites, for that matter) are very sound mechanical projections. They weren't built overnight, having been put there one microscopic bit at a time. One has only to attempt to break off a big one to find out just how strong they are. You can smash them with sledgehammers and do only minor damage. Even when water ceases to flow over their surfaces and the growth process is replaced with one of decay, little danger exists. They tend to wear away over *millions* of years in the same manner that they were formed—a microscopic bit at a time.

This is not to say that stalactites cannot or do not fall. Luray Caverns in Luray, Virginia, displays a massive column that was knocked over by one which fell into it. This obviously happened a few hundred thousand years ago and is quite rare indeed. This same cavern also has a very large room that is filled with thousands of stalactites, many of which are used to form the world famous Luray Caverns "Stalacpipe Organ" shown in Fig. 6-1. The organ console electronically controls electric strikers or solenoids which thump against a specific stalactite. Many are resonant at certain musical pitches, and enough different sizes are contained in this room to form a vast musical instrument which can play any tune a standard organ can. The constant thumping has not weakened these sturdy cave formations, which are subjected to a million times the vibration stress of a standard non-musical stalactite. This speaks well again of their mechanical stability.

As a teenager, I served as a tour guide at a local commercial cavern. One evening, a senior high school group was taken through, and some practical jokers ignited a cherry bomb in a room filled with hundreds of stalactites. The prank was pulled during a presentation in which all lights were out, so the guilty party or parties were never found. This is not the point, however. This tremendous explosion ripped through this large room and resounded throughout several others. Not one pebble fell from the ceiling, although many cavern guests (and their guide) found themselves on the floor due to a natural involuntary reaction. This may assuage the fears of those individuals who are scared to death to talk loudly in a cave for fear that the sound vibrations will bring the roof down upon them.

FLOODING

Another major fear is that of drowning due to the sudden rising of a stream in rainy weather. It is for this reason that it's not a good idea to explore caves which have or have been known to have a great deal of water in them during a heavy downpour or when a lot of rain has fallen over a short period of time. A few years back, a college spelunking team consisting of a professor and a

Fig. 6-1. The Stalacpipe Organ in Luray Caverns, Virginia, is made up of thousands of stalactites which are electronically controlled. (courtesy Luray Caverns)

fairly large number of students were temporarily trapped in a cave by rising water which covered the entrance. As I recall, a sudden downpour had created flash flooding throughout the area, so it may not have been possible to have foreseen this event. While this was an unfortunate happening, all safety precautions had been taken care of ahead of time, and when the group did not return on schedule, a team was sent out to handle the rescue. They were all found safe in a large room above the water line, although some were suffering from hypothermia, a condition brought on by heat loss. I have never heard of any natural cave filling completely with water due to a flash flood. Be careful, however, about entering caves with only one entrance that happens to lie beneath a *potential* high water line. It is possible for the entrance to become suddenly covered. From here on in, all you can do is wait to be rescued or camp underground until the water recedes. Incidents such as the one described with the college class are quite rare and should not be a cause of undue alarm.

LOST!

Another major fear is that of getting lost. This is a popular scenario in many books and movies. *Tom Sawyer* (the book and the movie) is an excellent example. While it is true that I have been turned around a few times in some of the larger caves, I have never been *truly* lost. Most caves have been explored to an extent that there are enough man-made markings to indicate the way out. These are the same types of markings that some spelunkers use to write their names on cave walls and are generated by the smudge from the flame of a carbide lamp held close to the surface of the wall. While I do not condone the latter, sometimes a helpful arrow smudged on the wall is a useful guide to finding a correct passageway. As you probably already know from the previous description of some of my personal cave exploring adventures, even the larger caves usually have only one or two obvious passageways. Many of the others may have hundreds of side passageways, but these quickly bottom out and one usually has to return to a main artery again. This is not always the case, however.

I have encountered some situations where I left some sort of marker to indicate the way out. This was not out of fear of becoming hopelessly lost, a situation I cannot conceive of, but to aid me in finding the right path and prevent unnecessary backtracking. And yes, some spelunkers do carry a ball of white twine that can be used for backtracking purposes in exiting a cave. I have never had to do this, but I always carry enough twine to be able to should the situation warrant it. (If you forget your twine, you can always drop bread crumbs.)

ANIMALS

Cave animals do not usually present any serious danger, although some can be bothersome in tight spaces. Bats generally tend to stay as far away from human beings as possible, and I have never been bitten nor heard of anyone who has, at least by a healthy bat. There is at least one cave in the western United States which has been known to contain rabid bats—or, more accurately, bats which carry rabies. These animals can be natural carriers of the disease without actually contracting it. In the early days of exploration, several people died from the disease. This danger is highly unusual in most wild caves, and the one mentioned contains millions of bats, along with more millions of nasty biting beetles that swarmed on the cave floor.

While not common in the eastern United States, some western caves (especially those near the Mexican border) can harbor millions upon millions of these flying rodents. While even great quantities of bats are not usually dangerous, their droppings can be. Called *guano*, tons of it are often mined for commercial uses. Guano is highly flammable, explosively so in large quantities. A tale was once related to me of a group of cave explorers who built a fire in the mouth of a cave rich in guano. The end result was that the whole top of the mountain was blown off and the spelunkers were never found. I cannot guarantee the authenticity of this story, but it does *sound* plausible. I have never explored a cave which contained tons of guano and don't believe I'd want to for aesthetic reasons as well as those of safety.

LIGHT FAILURE

The last fear we'll talk about is that of being lost in the dark due to the failure of a light source. This should never be a real danger, because safety rules in exploring caves dictate that each person carry on his body at least three different sources of light. The standard combination is the carbide lamp, a flashlight, and candles. I do not recommend flares because of the high quantity of noxious fumes they give off. The three-light-source rule also states that adequate provision be made to power all of these sources. For the combination given this would mean carbide, spare batteries, and ample matches contained in a waterproof housing. Also, with the carbide lamp, it is necessary to carry an adequate supply of water, especially when none is available in the cave. Candles are very difficult to work with in most caves due to the drafts, but they can get you out in an emergency. One of the members of my original three-man cave exploring club disregarded this three-light-source rule when he was showing a small cave to a friend of his. They went in armed with only carbide lamps and stayed long enough to use up their water supply. The cave contained no springs or streams, and it was necessary for them to feel their way out in the dark. Fortunately, they had realized their problem and were making their way back to the entrance when their last light failed. They had only to grope the last 25 feet to the entrance. If they'd been in a large cave, they could have had a serious problem.

Another rule was violated by this friend—and at the same time. The first light source rule goes hand in hand with another and is abbreviated by the saying, "Three light sources for three spelunkers." Each spelunker is to carry his own three sources of light, but there should always be a minimum of three working in a group. One never explores caves alone, nor with the aid of only one other person. *Three* is the minimum number for safe exploring, and this rule should never be broken.

The reason for a minimum of three people lies in common sense. If one person is injured in a cave, two persons may be able to get him out; one probably would not. Also, if the injury is great enough that movement of the victim is impossible or a greater health risk, one person may stay to attend to the injured party while the other can go for help. This rule applies for explorations of the smallest cave as much as it does to the largest. I have never been on an expedition where a serious injury occurred. However, I felt confident in the fact that if it *did* happen, the party was so constructed as to make immediate attention to an injury quite practical.

Hard hats should always be worn when cave exploring. No, they don't have to be of the variety shown in most articles on caving and in this book. While the front-brim construction hat is favored, one sees all sorts of arrangements, including war surplus GI helmets. The point here is to get something on top of your head that is harder than it is and which will tend to stay on.

This leads to some misconceptions, however, and I have seen more than a few amateur spelunkers who have attached chin straps to their hats. This is a safety hazard, not a help. While it is true that a hat will fall off occasionally, all that should be attached to it is a carbide lamp, and these are practically impossible to break. On one expedition, my hat became dislodged while climbing over a drop, and it fell over 60 feet, banging into walls several times on the way down until it reached the bottom. The hat wasn't even scratched up, and all that happened to the carbide lamp was a bent reflector. This was quickly straightened out and the expedition went on as usual. Why not put a chin strap on a hard hat to keep it from falling off? Because it is bigger than your head and you can get caught between rocks and other overhead obstructions and choke you. When your hat becomes lodged in an area that's just too tight, it should come off quickly to allow you to pull it free. A chin strap wouldn't prevent you from removing the hat quickly, but suppose you fell forward and your hat became lodged in a crevice? This is a good way to receive a sprained vertebrae or even to cause a break. When you're coming down a rope, both hands should be on it. If your hat gets lodged in a crevice, you can't take one hand away from the rope to pull it loose or you may fall. Don't plan to include a chin strap on any headgear you wear while cave exploring. Just make sure any hat you do wear fits snugly and will stay on.

Chapter 4 dealt with carbide lamps and how they operate. It was pointed out that the gasket between the carbide cup and the rest of the lamp must be kept clean and fresh to make a good seal. Remember, acetylene gas is forming in this area and will leak if the gasket is not fitted properly. Safety precautions require that you check and clean this gasket periodically or after every use. Several years ago, I was descending into Rodger's Cave (discussed in Chapter 5). I had noticed that my gasket was getting a bit worn, but a quick check with a match indicated there were no leaks. I was making my way down the second level drop (about 50 feet to the floor) when suddenly the entire cave seemed to light up. I hadn't realized what had happened, but my partners above screamed "Your hat's on fire!" I immediately knocked it off my head and it fell to the bottom of the drop. I was not burned but I was put in a very dangerous situation. I had been forced to take one hand off the rope while hanging in a potentially lethal spot, and after the hat was discarded I was without light. Naturally, I had my candles and my flashlight stashed in my knapsack, but I had no way to get to them and no way to hold them if I could. The only thing to do was climb back up again toward the light provided by my partners. When I was safely up, it was necessary to exit the cave and get a spare hat and carbide lamp before the expedition could be resumed. This resulted in an hour's delay, which is nothing compared to the potential danger I was in when the seal failed. Upon recovering the hat, it was discovered that the whole top portion had been smudged by flames which apparently encompassed the whole hat. If I had been situated in a downdraft, it's possible my face or ears could have been singed. My hair might even have caught fire. While this is a very unpleasant possibility, the fact that all this could have happened while I was suspended some 50 feet off the cave floor should be enough to make anyone check that gasket and make sure that it and the rest of the lamp are perfectly functional. Getting back to the no-chin-strap rule, you can easily see how this would have further complicated matters.

There is a great deal of ignorance regarding ropes. Many aspects of these spelunking tools were discussed in Chapter 4, but to digress, I feel that for serious climbing, one-inch, high-quality rope is mandatory. Short sections of safety line (half-inch manila) should always be used as well. These safety lines are attached to the waist of the climber to act as an emergency brake in case of a mishap. Sometimes it's not possible to use short lengths, especially when sheer drops are involved, but in any event, *all* safety lines should be discarded after each expedition and replaced with new ones. This may seem like a waste of money, and in many instances, it is. But remember, lives often depend on these ropes, and when used in emergency situations, they undergo tremendous amounts of stress. Ropes are subjected to a lot of natural abuse when used for caving. They become caked with mud and completely infiltrated with moisture almost immediately, and proper care (cleaning, drying, etc.) is mandated.

What about fires in caves? In some, small cooking fires can certainly be handled safely. In others, you're taking a very great risk. Many larger caves have good air circulation, as evidenced by a nearly constant breeze. Others, especially the smaller ones with tiny, single entrances, often have very poor ventilation. I am reminded of one of the first small caves I entered which contained a narrow ground-level entrance. Shortly after entering with two other people, we noticed that the flames from the carbide lamps began to dim. This was an immediate indication of an accelerated CO_2 content, and we immediately exited. The combination of three persons breathing the limited air supply plus the flame from the three carbide lamps had gone a long way toward depleting the small quantity of air which had gathered in this cave. It should be stressed that this was an extremely small cave, containing only one oblong room about four feet in height and ten feet long. Generally speaking, however, most caves will have good ventilation, but it is seldom necessary to build a fire in them, because most spelunking expeditions are simply not that long.

Every expedition should carry a first aid kit. What it contains will depend upon the paramedical expertise of the spelunkers. Obviously, it does no good to carry drugs and hypodermic needles if you have no knowledge of how to administer them. If you have paramedical training, however, and can legally administer certain drugs that are commonly used for emergency medical care, then by all means stock your kit accordingly. Most spelunkers, however, include common items in these kits—an assortment of bandages, adhesive bandages, splints, ointments, disinfectants, aspirin, smelling salts, and so on. It is nearly impossible to do any active cave exploring without sustaining assorted nicks and scrapes and a banged knee or two; these are the same afflictions common to those who indulge in any weekend sporting events, and certainly shouldn't be considered serious. Remember, though, that if you receive a minor open wound, it's best to clean it, treat it, and cover it *on the spot.* This is due to the fact that you're going to be crawling around in the mud, and there's a higher chance of infection. Normal scrapes and cuts that you might obtain on the surface would usually go untreated, but be more cautious underground.

Probably one of the most timely safety hints for spelunkers is, again, based on common sense. Never jump from one spot to another in a cave. Admittedly this rule must sometimes be circumvented to get from one point to another. The rule is a general one which is necessary because most cave floors are very uneven and are often strewn with a number of jutting rocks and other loose objects. It is sometimes very difficult to walk through level passageways because of the many loose obstructions encountered. One of my caving partners sprained an ankle during one expedition by simply dropping to the cave floor from a rope he was suspended on *one foot* above it. His boot landed on a roughly circular stone about an inch in diameter, and he lost his footing and went down. A corollary to this rule is to *always be sure of your footing.* Even the light given off by a properly functioning carbide lamp is quite directional, and you are generally required to examine the floor in front of you while it is illuminated, then to remember what was there, because when you walk over it your feet are going to be in the dark. It's quite easy to trip and fall, but fortunately, clothing can take care of many of the shocks and jolts.

For safety, cave clothing should provide several layers of protection while not being overly bulky nor of enough insulation content to make you perspire greatly. The "spelunkers's tuxedo" is a full-length coverall with long sleeves worn over standard clothing such as jeans and a flannel shirt. This provides two layers of protective clothing and will prevent many cuts and scrapes from acute stone projections. Again, the coverall should be almost form-fitting if possible. This type of garb is tradionally quite baggy, and it may be necessary to obtain the services of a seamstress to make for an ideal spelunker's fit. The coverall has been chosen as universal garb because it surrounds the body in a single package. Don't, however, choose the common type with a full-length zipper. Zippers and cave mud go together—but not well. A beginning spelunker has only to be cut out of his new pair of coveralls once to learn that a mud-infiltrated zipper simply won't budge. It used to be fairly easy to find coveralls which buttoned up the front. If you can't locate a pair in your area, have the seamstress remove the zipper and fit the open flap with four or five large buttons.

From an injury standpoint, knees are quite vulnerable. In many caves you use these parts of your anatomy more for walking than you do your feet. When climbing up or down, it's also easy to bang a knee into a limestone wall. This can create bruises that will keep you off future expeditions for the next few weeks. Rubber knee pads are available in many hardware stores and were originally developed for scrubwomen many decades ago who were constantly working on their knees. Most types have a hard rubber front pad that is ribbed. Straps and buckles are attached to the back and they are simply tied onto the knee joint. Other types are made completely of molded rubber; they simply slide over your feet and are pushed up to the knees. The molded rubber straps hold them in place. They look awful when strapped to the outside of coveralls, so if you're fashion-conscious, you can wear them underneath with the same results. (Of course, if you're that meticulous and fashion-conscious you won't want to be covered with mud from head to foot and probably won't go spelunking in the first place. Those opposed to dirt and grime will be anything but comfortable underground.

DRINKING WATER

Those of you who have traveled to Mexico and some other foreign countries can appreciate another safety rule for spelunkers, which simply states "Don't drink the water." This can be further explained by saying "Don't drink water that has not been properly treated." Nine times out of ten, underground streams are crystal-clear and *probably* safe to drink. Most of these are the result of subterranean outputs from underground springs. However, if there is a toxic waste dumping ground nearby, we all know that this can affect subterranean water. Also, rainwater flowing in from the outside can add to underground streams and pools and sometimes contaminate them. The point here is that *you just don't know*. For very long expeditions underground, some spelunkers elect to carry quinine pills which can successfully rid drinking water of certain contaminants. Alternately, the water can be boiled. Drinking water is usually very little problem, because most expeditions are of fairly short duration, and canteens can easily be strapped to a belt (worn inside the coveralls) or stowed in a knapsack.

Naturally, the water from underground streams can be used to fill carbide lamps, further preserving your drinking water supply. If you plan to use underground water for your carbide lamp, be sure to take along an appropriate scoop and funnel in order to get it into the holding chamber. During one expedition in Rodger's Cave, drinking water was accidentally left behind on the surface. This was not noticed until we were deep in the cave. Since we had intended to stay less than three hours, we were not concerned. However, our drinking water was also to be used to fill our carbide lamps. While we had two other sources of light (flashlights and candles), it's very difficult to climb a 75-foot drop with one hand on the rope and the other on the light source. We immediately beat a hasty retreat for the entrance, where a small stream was located. On the way back, however, our lamps began to dim. While we could have used flashlights to make it the rest of the way to the foot of the drop and the stream, it was still necessary to do a fair amount of crawling and climbing to get there, and we all needed both hands to accomplish this. There was only one way to keep our carbide lamps operational, so at a convenient point old carbide was dumped, fuel was replaced, and it was necessary to urinate into the top holding chamber of the lamps in order for the acetylene generation process to come about. This may sound quite comical, but it came about out of necessity and the act possibly saved one or more of us from taking a bad spill from having to hold onto a flashlight. You might credit us with resourcefulness at coming up with this idea, but if we had made a proper check before entering the cave, these unusual actions would not have been necessary. There was then and still is no excuse for entering a cave without the proper basic equipment. Naturally, when exploring a cave for the first time you often wish you had brought something along that you hadn't planned on needing. However, you will *always* need an adequate water supply and carbide, and to leave something of this importance behind is highly amateurish. While I plead guilty to this dubious title in the aforementioned case, let me defend myself by saying that it never happened again. Repetition of an amateurish act with such important ramifications is even worse.

This next safety rule should not even need to be mentioned, but unfortunately it is necessary due to the natures of some people. A cave is no place for alcoholic beverages. This applies to the period before entering the cave as well, and to at least sixteen hours before an expedition is to begin. Don't drink and drive, don't drink and fly, and for goodness sakes don't drink and spelunk. Sure, a cold beer may hit the spot when you exit the cave. Fine, if you don't have to drive back home, but even the presence of a small amount of alcohol in the body during an expedition can create a potentially hazardous situation. Think of it: you'll be operating at a fast exercise pace underground, consuming large quantities of oxygen and getting worn out at a rapid pace. These factors can lead to an advanced physiological impact of the alcohol in your bloodstream. While I have never seen anyone enter a cave in an alcohol-affected state, I have occasionally come across beer cans strewn haphazardly in the deep interiors of some systems. This would indicate that a spelunking party went in, did some exploring, then stopped for a few cool ones before making their way out. These morons didn't even have the courtesy to take their trash with them. To successfully and safely explore cavern systems, a cool, alert mind is necessary. Alcohol has no place in such environs.

General health is another factor. This has been lightly touched upon in a previous chapter, but a few more words are necessary. If you have heart trouble or a condition such as asthma which prevents you from strenuous exercise, then most forms of cave exploring are out of your physical reach; this is obvious. However, if you are just getting over the flu or some other temporarily disabling ailment, wait until you're completely healed and have your full strength back before taking part in an expedition. Even a bad cold can affect your performance underground, and the moist, cool air will probably aggravate the situation by lowering body resistance.

On a similar subject, if you have visual difficulties which require corrective lenses, take along an extra pair of glasses. I am extremely nearsighted and would be legally blind without my glasses in a cave illuminated only by the glow of a carbide lamp. I always carry an extra pair of glasses in an impact-resistant case as an emergency backup if my standard pair should be accidentally flung over some precipice. In 20 years I have always exited the cave with the same pair of glasses I wore in, but at any time this situation could change.

In many ways a safe subterranean expedition is like a manned space flight. In the latter, it's not possible to return immediately to home base to pick up something you forgot or to replace something that broke. In cave exploring, it may not be impossible to return immediately, but it is always impractical. Therefore, backup systems are mandatory. This is the reason for many of the previously discussed rules, such as three sources of light, adequate water, and so on.

Another major safety rule involves informing at least two responsible local people as to exactly which cave you're exploring, exactly where it is located, and when you expect to return. Additionally, they should be given an absolute time after which trouble can be assumed. In other words, you might tell someone on the surface that you expect to be out of the cave somewhere around 3:00 p.m., at which time you will give him a call. If they have not heard from you by 5:00 p.m., they should inform local authorities that there is a good possibility of a problem. Additionally, the name of the landowner should also be supplied, along with the type of authority which should be contacted. In most instances this will be a local police or sheriff's department. For goodness sake, be sure to communicate with your aboveground contact after you exit the cave to prevent the authorities from spending time and possibly taking personal risks in trying to get you out of a cave which you safely exited hours before. Always give this information to at least *two* persons, so one can act as a backup if the other is unable to assist due to some emergency.

I even go one step further by posting a notice at the cave opening stating who is inside, their telephone numbers and home addresses, the time the party entered the cave, and the time they expect to exit. This is especially useful should an accident occur in an area with many closely spaced cave openings.

Make sure your aboveground contacts understand the exit time fully. A confusion regarding this factor almost created an unpleasant incident when I was a teenager. A nighttime caving expedition was planned with an exit time of 11:00 p.m. maximum. One of the individuals was to be away on other business during the morning and afternoon hours of the same day and had neglected to tell his contact that it was to be a *nighttime* exploration. When he gave the exit time to that person on the day before the expedition, he simply stated that it would be "eleven o'clock." The contact assumed he meant 11:00 a.m. and was ready to call the authorities when he heard nothing. Fortunately he called another contact first, who explained that the group hadn't even entered the cave yet.

In addition to protecting yourself while in the cave, it is also necessary to obtain some medical attention aboveground before actually entering. It has already been pointed out that nicks, scrapes, and cuts to the skin are an occupational hazard of cave exploring. When a cut occurs, if not treated immediately it is exposed to cave mud which often contains bacteria that are potentially dangerous. Often the bacteria may be present on the rock wall or floor which causes the break in the skin, so the bacteria enters immediately.

The bacteria *Clostridium tetnae* is in all types of soil, both above and below ground, and is especially prevalent in areas where animals have been. The only way it can enter the body is through a break in the skin. This bacteria is often called rusty nail syndrome, as it is the cause of tetanus or "lockjaw," as it is sometimes called. For this reason it is imperative that all potential cave explorers receive the required tetanus-toxoid injections for immunity to the potentially fatal outcome of this type of infection. Most of us in the United States receive these basic injections before entering grade school. However, a tetanus booster is often required when a severe skin puncture occurs. Since there is a chance that this bacteria will be found in larger quantities in some caves, a booster shot should be given before entering your first cave, and then once every year thereafter. Check with your family doctor and ask what he recommends. Be sure to explain that you intend to become active in cave exploring and outline the fact that this bacteria has been known to inhabit subterranean passageways in high numbers.

LEADERSHIP

Finally, for an expedition to be truly safe, it is necessary to appoint a leader. It is then mandatory that you go one step further and make sure everyone understands that what the leader says will be immediately obeyed without question. The good leader will ask for input from his fellow spelunkers, and if his performance is not satisfactory, another one can be chosen . . . *after* you're out of the cave. Leadership is not an easy job, and it's quite easy to become unpopular. However, only those who have been leaders of caving expeditions realize fully the responsibility which is placed upon them. As far as I'm concerned, if an argument breaks out between expedition members over whether or not to follow the leader, that is immediately justifiable cause for scratching the entire expedition and heading back to the surface. While the leader's responsibility is multiplied tenfold when an accident or other emergency situation occurs, his real responsibility is in doing his utmost to keep these things from happening. Naturally, in a larger expedition (six or more persons), an assistant leader is necessary to take over should a situation occur which render the leader incapable of performing his task. Although expedition members often gain experience together, all having started as rank amateurs, eventually a new member or several may be added. Here is where discipline often breaks down, in the excitement of one's first trek in an underground cavern. This concept may sound ultra-military, but organization is an absolute necessity. (This pertains to equipment as well as to spelunkers.) Again, *before* planning any expedition and actually entering a cave, elect a leader and then obey his every command, even if you don't happen to agree with it. Leaders are not perfect and they do make mistakes. These mistakes, however, are discussed only after the expedition is completed, and preferably no sooner than the next day. If the precaution of electing a leader is not taken ahead of time, then all other safety measures, though followed religiously, may be to no avail. After an expedition, the performance of every individual in the expedition from the leader on down to the greenest spelunker should be evaluated. This is especially necessary when further expeditions are planned with the same group. If orders are not being obeyed, the person causing the problem should be alerted to this, and if the situation doesn't improve, you'd be doing him a favor and yourself as well by excluding him from further explorations. Again, the concept of performance applies to the leader as well as to those who have promised to follow him. If any accidents occur to the expedition party, it is the leader's responsibility. While an accident may not be his fault, he is still responsible for it and must take steps necessary to treat the victim and bring him to the surface again.

This last safety precaution should come first.

Unfortunately, to many expeditions, it is not even considered. What results is an unorganized group of individuals who are just out to have fun and don't realize that this is often impossible without proper leadership and organization. The individual who strikes out on his own without regard for his safety or that of any other expedition member jeopardizes everyone. He could fall and break a leg; and although he has brought this on himself personally, it then becomes the responsibility of everyone else to get him to safety. An underground rescue attempt increases the possibility of subsequent injuries to the rescuers.

Robert A. Heinlein, an author since the 1940s and an extremely remarkable individual (judging by his books alone), told a fascinating story in a recent book entitled *The Number of the Beast*. It involved a four-person expedition (two married couples) and the leadership difficulties they encountered. The original leader became frustrated with the obvious lack of obedience and resigned. Another leader was appointed, and she in turn resigned. The leadership role went full circle and finally, after everyone had had a chance and fully realized the responsibility and extreme difficulties of leadership, the group elected a permanent leader. While this was a book of speculative fiction, Mr. Heinlein has pointed out time and again that all fiction must have a basis in reality and authenticity. I suggest to you that this procedure might be a good one to follow in selecting a permanent leader from a group of friends who wish to form a cave exploring club. In *The Number of the Beast*, leadership problems were complicated by the fact that all involved were close friends, and this is the most difficult type of organization from which to obtain the discipline that leadership requires. This is also often the case with groups of people who explore caves.

Use any system you deem appropriate to choose a leader. Just make sure that this matter is taken care of. I think you will find that your experiences underground will be far more pleasant, certainly safer, and will continue year after year.

CAVE COURTESY

Cave courtesy simply means thinking of your fellow man and the environment that you have entered. Like many safety rules, cave courtesy is self-evident and based upon common-sense judgment. Certainly, it's in bad taste to enter a beautiful underground chamber and leave behind scraps of paper and soft drink cans or childishly write your name on the wall. This is something you certainly wouldn't do in your own living room, or anyone else's for that matter. The beauty of caves and cave exploring lies in the fact that you see and experience something which the average individual has not. You're actually going back in time when viewing an underground panorama that is almost exactly like it was thousands of years ago. A cave, then, is a time machine into the distant past when the earth was still young. Perhaps a thousand years from now, someone else will see those sights exactly as you see them. I have an almost religious aspect for what is contained in wild caves and I always attempt to leave the underground intact as if I had never been there. If you have never been in a wild cave, then you may not be able to appreciate the deep feeling for them that we who explore the underground cherish. When exploring a cave, take out with you everything you took in; disturb nothing and leave only footprints behind.

Cave courtesy extends aboveground as well. Remember, that cave entrance lies on someone's property and permission must be obtained before you enter. Often, caves will be located on farmland, so you don't want to drive your vehicle across a farmer's garden, scare his livestock to death, etc. Sometimes it's necessary to travel through gated fences, and naturally, you should always leave the gate exactly as you found it. If the gate is open when you arrive, don't feel you'll be doing anyone a favor by closing it after you pass through. It may have been left open to allow livestock to enter from another field. Likewise, if the gate is closed, open it, pass on through and then close it again securely. By exhibiting the proper behavior aboveground, you're not spoiling your chances of returning again for another expedition in the future. Also, the landowner will be more inclined to permit access to his cave by other expeditions.

Respect the cave flora and fauna as well. Don't

leave pieces of twine or other similarly small objects behind. Bats and other cave animals can become hopelessly entangled and die horribly. Don't even toss scraps of food around in the hopes that they may be devoured by some form of underground life. If everyone did this the bottoms of our cave floors would be one big, rotting garbage dump. Don't unnecessarily frighten bats and other cave creatures. This is *their* home, not yours. And again, don't dump your carbide haphazardly. Stash it in a bag or container and take it out with you for deposit into a proper trash receptacle.

As another courtesy to those who will follow you in exploring a particular cave, make note of any unusual and potentially dangerous situations that you happen to find. For example, if a drop occurs suddenly in a passageway and is not readily recognizable until the last second or so, you might smudge the phrase "DROP AHEAD" onto the cave floor or wall. I try to do this in mud banks or other areas where the smudge is erasable. Even though the mud is constantly damp, it does not do a great deal of traveling, and that message could remain there for many years. You might also attempt to rough out a crude map of the cave, pointing out any unusual situations and indicating the best passageways to take. This could be left with the landowner, who might like to pass it along to future expeditions.

Also make note of conditions of which the landowner should be apprised. For example, if you find signs of livestock entering a cave opening with the potential of being killed or injured, you might want to mention this so that steps can be taken to fence them off from the entrance. Also tell him if there is evidence that previous explorers have been abusing the cave, and try to carry out any debris someone else may have left behind. If you have permission to camp overnight on another person's property in order to explore a cave the following day, make sure your campfires are kept in check and that the entire area is policed before going into the entrance. After you come out, you will probably be too tired to do this efficiently. The owner certainly won't appreciate it if you leave trash aboveground any more than below.

Be sure to obtain *written* permission and have that important document with you at all times. This is especially necessary when the land is posted to trespassing. In some states, non-posted land requires only verbal permission, but those properties which are posted require it in writing by law. Even though for a previous expedition you may have been given written permission, if you go back again, always recheck with the landowner to make sure it's all right.

Cave safety and courtesy are absolutely essential to the continuation of underground exploration. More than one cave has been put off limits to all explorers due to the childish and often dangerous actions of a few. Remember, if your act is not as clean as it could be, you may not only jeopardize future explorations for yourself but for the rest of us as well.

As a final act of courtesy, on the way home after an expedition be sure to stop and thank the landowner for allowing you to visit on his property. This is always appreciated and will better your chances of being able to come back again and again.

CAVING ACCIDENTS

With the proper blending of experience, knowledge, caution, and maturity, spelunking is a very safe pursuit. However, accidents can occur even when precautions are taken. Most of the time, though, some standard safety rule is violated, and this is where people get into trouble. The National Speleological Society publishes a booklet entitled *American Caving Accidents* which is excellent reading for anyone presently involved in or considering cave exploring. In this booklet are brief descriptions of every reported caving accident occurring in the United States, Canada, British Columbia, and Mexico, usually or a four-year period. As a further help, there is an analysis of the accidents with an opinion of why it happened and how it could possibly have been avoided. This booklet considers an accident to be any situation where an injury occurs in a cave, or when a spelunker gets into a situation where the immediate assistance of others is needed.

Who has caving accidents? That's a difficult

question to answer, because anyone who takes up cave exploring is a potential accident victim. (Naturally, the same can be said of any person who drives an automobile, engages in a contact sport, etc.) A recent report shows that the age of the largest group of accident victims is between 16 and 20. As a matter of fact, this group, spanning four years, has about as many accidents as are reported in the 21-to-30 age group. These figures may be partially attributed to the fact that many individuals take up cave exploring between the ages of 16 and 20, when inexperience is often coupled with less-than-fully-developed maturity, and situations develop in which accidents can occur.

Notice that the previous statements only *partially* attributed accidents to inexperience. Cavers with only a moderate amount of experience have the least accidents. Those with little or no experience have about three times as many accidents. Surprisingly (to some, but not to others), the group which has the most accidents are the *experienced* spelunkers, coming in at about 33% higher (in accidents) than a group with little or no experience! Of course, the experienced cave explorers probably spend more hours in caves each year than do either of the other groups, but this *should* be offset by the increased experience level.

That experienced cavers seem to be more prone to accidents than the other groups mentioned comes as no real surprise to me. It's also a fact that experienced electronics technicians are more apt to be injured or killed by coming in contact with high voltage than are those with moderate to little experience. I'm sure similar comparisons can be made in other professions and pursuits.

In potentially dangerous environments, familiarity does not breed contempt; rather, it breeds carelessness. When you first descend a 100-foot drop in a cave, chances are you are extremely cautious because you're scared. This is a natural human reaction, and is why we stay alive and healthy. However, when you have descended a fair number of drops without incidence or accident, it is only human nature to assume that you are fully capable of continuing these experiences in complete safety. This is basically true . . . but *only as long as you remain in a highly cautious frame of mind*. When you take a sheer drop lightly, you are more susceptible to falling down that sheer drop.

During my early caving days, I witnessed an experienced spelunker climbing up a 70-foot drop without the direct aid of a rope. True, the rope was nearby, and he *probably* could have grabbed it if he got into trouble . . . but why take chances? It would be another matter altogether if the rope had broken before the ascent and this sort of practice was mandatory. (In most cases, this type of situation would be caused by poor planning or improper safety precautions.)

It can be truthfully said that whenever someone takes safety precautions lightly, he is several hundred times more likely to have an accident, regardless of his relative experience level. Throughout this book, safety precautions are mentioned again and again, but there will be some people who will eventually bypass one rule (or several) for the sake of apparent convenience. *Don't do it!*

In looking over accident reports, it would appear that party size (the number of spelunkers on a single expedition) plays some part. The highest number of accidents usually occurs to parties of two to three. It is necessary to break these figures down a bit. Three should be the absolute minimum of any caving party, and this happens to be the standard number of spelunkers on the majority of expeditions, so this may be one reason why accidents occur in parties of this size. There are simply many more parties of three than any other number. Knowing this, it would probably be safe to say that the highest accident rate (*relatively* speaking) occurs to parties of two. There are far fewer parties of this size actively spelunking, but their accident rate is about equal to those involved in parties of three, which are far more numerous.

As party size increases, accident rates seem to taper off. This is understandable in that there are more people available to assist spelunkers who may not be all that experienced. This assumes that there are enough experienced spelunkers to keep an eye on the inexperienced members of the group. When the proper ratio of experienced and non-experienced persons in a single party is not maintained, however, the accident rate goes back up again.

The single most common direct cause of accidents or bad situations in caves is falls, and the single most common direct reason for a fall involves the spelunker trying to do something that exceeds his ability. Accidents attributed to this are about three times as high as those falls caused by equipment failure, inexperience, etc. Other significant accident causes involve disorientation, flooding of the cave, fatigue, and hypothermia. The latter has been discussed previously but bears mentioning again. When your body temperature begins to drop due to cold underground temperatures, your performance can be adversely affected to a high degree *without you being immediately aware of it*. You don't want to dress so warmly that you perspire (perspiration on the skin can cause the body to cool rapidly), nor so lightly that your body cannot retain heat.

The highest contributory cause of the accidents mentioned is poor judgment. This simply means that the caver was not experienced enough to make the correct decision, or if experienced, disregarded a safety rule. Following closely behind poor judgment is poor equipment and lack of preparation. Any one of these situations can breed a nasty accident. When you combine them, however, your chances for a bad experience increase rapidly. Anyone who enters a cave poorly equipped obviously exhibits a lack of preparation and is guilty of poor judgment. It is not unusual that poor judgment is the major contributory cause of cave-exploring accidents; this same factor is the major reason for automobile accidents, airplane crashes, etc. Again, caving is a relatively safe sport . . . it is, however, very unforgiving of mistakes.

Another report of caving accidents gave listings broken down by the month of the year. While more accidents seem to occur in the warm months, this can be expected since this is the prime time of year for spelunking. It should be noted, however, that there is a high incidence of accidents in the months of March and early April. In many areas of the country, these are the times when snow has left the ground and warmer weather is beginning to make itself known. While this is a pleasant time for those who remain aboveground, it is also a period when changes are occurring within the underground (the spring thaw, it's often called). During periods of thaw, there is a greater tendency for the upper surface of the earth to shift slightly (due to expansion), bringing about other shifts further underground. This is the time of year when underground slides and collapses would be most likely to occur. Even then, these events are very rare. As a matter of fact, collapses make up an extremely small percentage of underground accidents.

The main thing to worry about during the early spring and late winter periods is rapidly changing weather conditions. Torrential downpours are not uncommon, and low-lying cave entrances can sometimes be flooded. Occasionally, one hears of a caving expedition trapped underground by rising waters. In the great majority of such instances, the spelunkers head to a high part of the cave and wait for the waters to recede, or for a rescue party. (Of course, the latter depends upon preparation. If no one knows you're in a cave, a rescue party may never come.) Entrapment is a potentially lethal occurrence. On a few occasions, people have died from exposure due to such entrapment from rising waters. There is a good chance they could have been brought out safely had they been properly dressed and equipped with some high-caloric food items. Even if you're planning a brief stay underground, it never hurts to pack an assortment of chocolate candy bars and a few other food items. During an emergency situation, these can sustain you for surprisingly long periods of time and can mean the difference between life and death.

Some incidents simply cannot be anticipated when cave exploring. It's always possible to sprain an ankle or even break a bone, even when the utmost in precautionary procedures are practiced to the letter. Here, pre-planning has a lot to do with the ultimate severity of such an accident. If you're warmly dressed, have at least two other explorers with you, and have a food supply which will get you by, then you can spend quite a few hours or even a day or so underground while waiting for assistance.

However, if you have entered a cave unprepared, such minor accidents can quickly become major.

Accident reports are well-stocked with cases where a single individual or a small group interested in cave exploring encounter a cave opening quite by accident while pursuing other interests. Having been in caves before, they foolishly enter "just for a quick look-see" before coming back another day equipped for a true expedition. Unfortunately, during this first pre-expeditionary entrance, none of the individuals is equipped with such basic items as carbide lamps, coveralls, hard hats, etc. Often in short-sleeved shirts and with a weak flashlight or two, they venture into the opening. Finding something highly interesting, they push onward and someone suffers an apparently minor mishap. With flashlights failing and bodies rapidly losing heat, a highly serious accident occurs.

One such case involved three individuals, two weak flashlights, and little else. Well into the cave their first light failed completely and they attempted to make it out with the everweakening light that remained. That too failed long before they exited, and for hours they attempted to find their way out in the dark. No one knew they were in the cave, and although they were reported missing, the authorities made the standard aboveground checks. This story does have a happy ending, however. Due to a great streak of luck, the same cave was to be explored by a well-equipped expedition two days after the now-lost party entered. The second expedition came across the first and led them to the surface. This occurred in the late winter months, so the first party was fortunately well-dressed for the low outside temperature. They did have a few candy bars with them, so between these and a lot of exercise to keep warm, they suffered no long-lasting physical effects from the experience. Luck alone saved them, however. Had the second expedition never entered, it is almost certain that all three would have perished.

Equipment failure or improper equipment is another cause of many caving incidents and accidents. I'm not referring here to scuba gear, specialized climbing rigs and the like, but rather to the basic equipment such as hard hat, carbide lamp, etc. As discussed in Chapter 4, hard hats should be strapless. When you tie that hat to your head with a chin strap, you're setting yourself up for possible strangulation. This occurred to one caver in the late '70s when he suffered a short fall while climbing. He fell less than a foot, but his helmet became wedged in a crevice and he was left dangling in free space. Before the remainder of his party could assist him, he had suffocated from the pressure of his neck against the strap.

Some cavers use a specialized strap known as a chin cup arrangement. This is popular with bicyclists and other individuals who wear hard hats and need to keep them firmly in place. Instead of attaching at the back of the chin around the neck, the two straps terminate in a form-fitting cup which closely hugs the front of the chin. These devices will not strangle you should your hat become wedged, but I personally don't care for them because there's always the possibility that the pressure of a sudden fall could severely injure or break the neck. Remember, the hat is slightly wider than your head. If you're traveling forward in a passageway and the hat becomes wedged in an overhead crevice, your forward motion could cause a serious neck injury. There's a good chance that the strap would unsnap and free you, but you just never know.

There are some who would say, when discussing the chin strap fatality just mentioned, that if the hat had not wedged in the crevice, the spelunker might have fallen to his death anyway. This is probably not true. This unfortunate victim did not release the rope until after he had fallen and began to struggle. If he had not worn a chin strap, he probably would have received no more than a bump on the head. The neck is not intended to support the entire weight of the human body. It is my recommendation that no strap whatsoever be worn on any hard hat used for cave exploring. There are those who will argue that the chin strap is perfectly all right and even recommended from a safety standpoint because it does assure that the hat will stay on . . . this being safer than the hat coming off and exposing the head. While it is true that strapless hard hats will sometimes come off, this usually only occurs when it receives a severe blow. Naturally, you will want to make certain that the headband is adjusted to assure a firm, comfortable fit.

While not a factor with most male spelunkers two decades ago, today hair length can play a role in accidents, especially when climbing. A few incidents have been reported of cavers with long hair becoming entangled in ropes while ascending or descending a drop. Play it safe; if your hair is of sufficient length to substantially exit the rim of the hat, it should be tied off and either shoved up under the hat or down the back of your coveralls. Do not allow it to be exposed to possible entanglement or to the flame of your carbide lamp or those of others in your expedition. Nothing complicates a steep descent more than having your hair on fire.

While coveralls are standard attire for cave explorers and have been discussed elsewhere in this book, a fairly large number of people still explore in blue jeans and flannel shirts. I have done this myself on occasion, but I'm not particularly proud of the fact. When climbing, it's quite easy for belt loops, belts, sleeve cuffs, etc., to become snagged. This is far less likely to occur when wearing the standard "spelunker's tuxedo." Coveralls serve to more or less streamline the body (even with their baggy appearance) when sliding through uneven fissures. If you can get away without wearing a belt around the waist of the slacks you wear beneath the coveralls, this is recommended, because the underclothing can sometimes hang up even with coveralls.

While spelunking has long been a manly art, more and more women are pursuing this sport every day. Due to the differences in the female physique, some special recommendations are in order. Coveralls should be *mandatory*. There have been incidences where breasts have become entangled in ropes and other climbing equipment. A bra has the same effect as a belt, and is subject to snagging when crawling through tight passageways or descending into narrow pits. Unlike belts, however, when bras become snagged, arm movement is often restricted. Given the right set of highly unusual circumstances, they could even be the cause of strangulation. When cave exploring, it might be best to go braless or wear several layers of upper body clothing. Make certain, however, that you do not dress so warmly as to perspire.

The same recommendations and precautionary routine for men with long hair applies equally to women. Naturally, all rings, earrings, jewelry, etc., must be removed before entering a cave. This applies equally to men.

WHAT TO DO IF . . .

One of the many aspects which leads people to explore caves is the challenge of the unknown. Sure, you can anticipate most of the general things you will experience, but there are always unknown factors that cannot be directly anticipated but which proper pre-planning and safety precautions do take into account. Cave exploring is a contact sport, although you'll be coming in contact with hard stone instead of the relatively soft human body. No one has been a spelunker for a number of years without occasionally banging a kneecap, spraining an arm or leg, or inflicting a minor cut or two. These things are all part of the sport. You can eliminate over 95% of the potential for serious incidents and accidents, however, by following the rules of the NSS and proceeding with required caution.

Since there are potential dangers in the art of spelunking, it never hurts to think ahead and even practice what you would do if a certain emergency situation developed. We will attempt to explore a few possibilities in this discussion which may help you in the years ahead. The following materials include some emergency situations which have occurred to spelunkers in the past.

Let's assume that you are deep within a large cave and suddenly find yourself lost and unable to find any conventional points of reference. What do you do now? If you're like some, the first thing will be to dash about madly in an attempt to locate some passageway, formation, or subterranean trait that you can remember. If you do this, however, there's a good chance that you will become more hopelessly confused. The first thing to do in a situation like this is to sit down and try to relax for a few minutes. Then discuss every aspect of your exploration to this point with your fellow spelunkers. Try to recreate a mental picture of every move that you've made since entering the cave. It may be necessary to close your eyes and visualize every left-hand turn, every right-hand turn, each climb, each time you had to crawl, etc. Obviously, one doesn't become lost from the start, so it will probably be necessary only to think back to your moves within the last half hour. With input from your fellow explorers, there is an excellent chance that you will be able to decipher the moves your party made up to the point where you became disoriented. This method works well if you do it *before* you panic. If you madly dash around upon discovering you are lost, it may be impossible for you to remember all of these most recent maneuvers. The rule here is: When you realize you are lost, *don't take another step*.

Through reasoning and discussion with fellow spelunkers, you will probably come up with a good idea of where to start looking for the way out. *Never* separate the group under these circumstances; continue as before. Before striking out again, mark the spot where you first discovered you were lost. If your reasoning has not yet uncovered the way out, you can return to this point and try again. When moving away from this spot, leave markers every fifteen feet or so to guide you back.

Since the cave floor is often covered with damp mud, you may be able to locate your footprints and simply follow them in reverse order. It shouldn't take you very long to return to an area of the cave which includes a main passageway or some familiar point to get you to the surface. This is all that is usually required, as in most instances, we're talking about a minor disorientation rather than being hopelessly lost.

Let's carry this example one step further and say that after a half hour or so, you are still lost and all the reasoning power you can muster just can't seem to get you to a familiar passageway. The thing to do here is return to the previous spot where you thought about your situation and make further plans. If you're at all human, you will probably be getting a bit concerned by now, but if you think about your overall situation, you really should have nothing to worry about. This assumes that you followed NSS rules and practiced correct preplanning—you have informed at least one person on the outside as to your location and the time you expected to exit. If worse comes to worst and you cannot find your way out, someone will certainly come looking for you. Those food rations you packed in your kit might come in handy at this point. Great concern often saps body strength, and a chocolate candy bar, eaten slowly, can often bring about renewed physical strength and a better mental outlook.

At this stage, give some consideration to your light sources. Do you have enough carbide on hand to see you through an extra ten hours underground? Even if you do, don't waste your limited illumination. When not actively searching for a way out, it might be best to burn only one carbide or electric lamp at a time. Remember, if someone knows where you are, and you and the other members of your party are in good physical health, there is really nothing to become unduly alarmed about. The worst thing that can happen is enduring the embarassment often involved in being hauled out by a rescue party and the subsequent newspaper stories that follow, often headlined "Cavers Hopelessly Lost Underground." If you have a good light source, all party members are healthy, and panic has not set in, then you can begin looking, more or less at random, for a passageway or point that you may have overlooked before. Again, this is always done with the entire group (if small) or with at least two other members if the group is large. There should always be three persons together in *each* group at all times. You plan this search to extend outward from your central point like the spokes of a wheel. Mark each area explored every time by piling up stones or in some other manner to prevent you from searching areas you have already explored. Using this process, you will almost certainly find your way out of the cave.

The fact of the matter is that most wild caves are simply not so large as to consume hour upon hour of exploring time. I have been disoriented a time or two in large caves, but never for longer than about 15 minutes. While being lost is a major fear of many who know little about cave exploring, it rarely happens in actual practice.

One situation which can keep you trapped in a cave and unable to find your way out is to lose all illumination sources. This is an extremely rare occurrence, but can happen if the required three light sources per man are not taken into the cave or if a great deal of equipment is accidentally lost. Again, it is necessary to sit down, relax as much as possible, and assess your situation. Let's assume that you have lost all light and are at a known point in the cave but in total darkness. Depending on the particular cave it may be possible to "feel your way out," but probably the best thing to do is just sit calmly in place and wait to be rescued. If you are in a main passageway that has no drops or side channels which you could enter and become even more disoriented, there is a good chance you can crawl out in single file, with each spelunker holding onto the one in front of him. If there is the slightest chance of encountering a drop, however, just wait in place until someone comes looking for you.

Here, telling someone on the surface when you will exit the cave and what to do if you don't is of *prime* importance. If no one knows you're in the cave and you have reason to suspect that no one will eventually assume that you are there, then you have no choice but to try to feel your way out. Sometimes torches can be made from strips of clothing wrapped around a wooden object, but this makes for quite difficult going. *Be certain you tell someone on the surface, and preferably two or more people, of your planned exit time and exactly what to do if you do not report in by this time.*

Let's take another hypothetical situation. Three spelunkers are in the depths of a cave when one suddenly falls and breaks a leg. The first thing to do is tend to the injured party. If it is a compound fracture, it will be necessary to stop the bleeding and secure the injured leg. Then one person must go for help while the other stays with the victim. This actually breaks a rule of cave exploring in that the group is split up into parties of less than three persons, but during an emergency, this cannot be helped. The person remaining with the victim will tend to his immediate needs and keep morale up. It may be some time before a rescue party can be summoned. The spelunker who goes for help must be able to lead rescuers back to the injured party and explain what may be encountered in getting him out of the cave. This assures that the rescue party brings along the equipment needed to facilitate a speedy and safe rescue.

In most instances, it's not appropriate to remove the victim yourself. In the case of a leg injury, it would be necessary for the two healthy spelunkers to carry the injured party out. This might be practical *if* there are no drops to negotiate and *if* the passageways are fairly large and level, but it is quite easy to become exhausted in an underground environment. This can occur quite suddenly and could put the formerly healthy spelunkers on the disabled list, making them too tired to exit the cave. Now no one is left to go for help. If the leg injury is a minor sprain, it might be possible for the disabled spelunker to limp along and help with his own rescue. This is where common sense is called for. Some injuries will not require a rescue party, while others will. In any event, don't hurry. Tend to the injured party and then discuss your plight and determine which procedure should be followed. The victim may be able to indicate whether or not he feels he can exit the cave with aid. Depending on the extent of the injury, his advice may or may not be a source of help.

Taking another hypothetical situation, let's assume that you and two other spelunkers are crawling through a narrow passageway and one becomes stuck. This is quite a frequent occurrence, but most stuck parties can extricate themselves with a little effort. Sometimes, however, assistance is required. There will be a tendency on the part of the individual who is temporarily held in place to panic. He may twist and turn and become even more tightly wedged in place, or he may injure himself, causing body swelling. Either of these situations is not conducive to an easy removal and makes the situation worse.

Often, when people become wedged in tight places, it's because of a piece of clothing or equipment that's become hung up. Examine the individual carefully and see if you can locate any items that might be removed to make extrication easier. It will probably be necessary to calm the individual down, especially if he's new to spelunking, and then proceed quickly, but not at such a pace that mistakes can be made. If removal of gear does not effect his release, try smearing the wettest cave mud you can find over his coveralls and body. It may even be necessary to cut away some clothing, but don't remove too much. If a rescue party must be called in, the stuck individual might suffer from hypothermia if forced to wait unclad for a long period of time. With the slimy mud in place, some forceful tugging on arms or legs may result in a speedy removal.

When a person becomes wedged in a passageway, it is usually a small problem to remove him. But should a rock shift or a collapse occur, the removal becomes highly complicated. If the individual is injured, he must be treated as best you can immediately. You can then look over the situation and estimate whether or not you have the resources to free him. Be careful about digging around in the loose debris which often makes up a collapse; you could bring about an even larger collapse that might also trap you. If an injury has occurred, it is best to go for help immediately, especially if the shift or collapse has stabilized and there is no immediate danger that the situation will become more complicated. The party that goes for help will again need to lead the rescue party back, but will also need to explain in detail just what has occurred and what might be required to remove the trapped individual safely. The other member of the party remains with the victim, keeping his morale up and doing his best to extricate him if possible.

As another example, let's assume that a party of eight has entered a cave and one spelunker has become separated from the main group and is apparently lost. The first thing to do is to keep the main group together. Do not allow individuals to go dashing off in all directions. As in a previous example, everyone should sit down and go through the exploration to this point in as much detail as possible. It is necessary to establish when the missing individual was last seen. Certainly, you will want to try calling out in hopes that the missing person will hear you. If you do make communication, encourage him to remain in place, as it is sometimes difficult to tell where voices are coming from when underground. Since we're talking here about a party of eight with one separated from the other seven, the procedure is quite simple. Four of the group should remain in place and continue calling out at regular intervals. A group of three will then backtrack in an attempt to locate the missing person. Using this method, the individual is usually found quite quickly. One of the three members of the scouting party can then make his way back to the waiting group of four to bring them on to the site where the missing party has been found if their assistance is needed.

If *you* should be the missing party, your best bet is to remain in place and call out at regular intervals. When you first become aware that you are separated you shouldn't be that far away from the path that was taken in the exploration of the point where you became separated. If you wander around trying to locate the rest of the party, there is a chance that you will move deeper into the cave and into an area which is farther away from the rest of your group. When larger parties are properly planned, someone experienced is always at the rear to watch out for strays, so this type of situation is quite rare, but it has occurred on occasion. If you panic, all is lost; if you remain calm, there is a good chance that a potentially serious situation will turn out to be only a minor incident.

Let's assume a worst-case situation in which someone becomes seriously injured during an exploration and there is really no time to go for help. It then becomes the responsibility of the healthy members to remove that person from the cave and drive him to the nearest hospital. If passageway size permits, you can sometimes fashion a crude litter from coveralls that have been split down the front. This can be used to drag the individual or carry him to the surface. If drops are involved, you will have to secure the victim as best you can to haul him upward or lower him. In this type of emergency, you do the best you can. Do not take chances with your own safety; if you become injured, all may be completely lost. Sometimes you have no choice but to leave the victim in the care of a healthy spelunker and do your best to recruit local help for his removal. Get him out if you can; if you can't, go for help, even if his needs are immediate. Local residents may know of a nearby doctor or other medical personnel who may be drafted into entering the cave with you to provide emergency care on the scene.

In summary, most incidents and accidents which occur underground can be avoided by following NSS safety regulations to the letter. Should a freak accident occur, a cool head will generally see you through. When you meet with other spelunkers, discuss possible emergency situations and what you would do if faced with handling them. When practiced as a group discussion, there will be many different ideas on ways to handle emergencies. Many of these may potentially be put to practical application should an accident occur. The National Speleological Society can provide you with a lot of information in this area, and their excellent publication *American Caving Accidents* should be required reading. If you belong to a cave club, it might be a good idea to have a member of your local rescue squad speak before the group, and all members should be required to take EMT courses which are often made available free of charge.

Chapter 7

Cave Exploring Clubs and Groups

Safe exploration of caves mandates parties of three or more individuals. This is all that is required to remain within the safety rules, but a party of this size is sometimes limited in what it can accomplish in a large cave. Then too, it is difficult for such a small group to expand interests quickly and become more proficient at the many side aspects of this exciting sport. Three individuals just don't offer a broad enough range of interests and expertise.

Those who become really interested in spelunking often attempt to expand the number of individuals in order to bring in more ideas and talents. When ten or more people are involved, many more options will be expressed, and interesting ideas can certainly grow from these opinions. This is not to say that all ten will necessarily be included on every underground expedition. While this may occur occasionally, the main idea here is to obtain assistance from a sizable number of individuals, each of whom may be involved in a different profession, offering expertise in different fields.

You will find that some individuals take readily to climbing and general rope work, while others need much more training in this particular area. Some will be quite handy with gadgets and can repair equipment or even design new types based upon a specific need. A lawyer, for example, might be in a better position to arrange for permissions from landowners and can provide the proper legal assurances regarding release from liability. A long-time outdoorsman who is accustomed to camping, hiking, and so on will be able to give advice concerning the equipment needed for extended expeditions, footwear, etc., based on past experience. A doctor who is interested in spelunking can add greatly to cave safety by advising members as to emergency medical procedures and provide direct assistance should a problem develop during an expedition of which he is a member.

Using these professions in pursuits as examples, you can probably get a pretty good idea of how to proceed in the forming of a cohesive group of individuals who devote at least a portion of their spare time to cave exploring. Those who display expertise in one field serve as excellent teachers and can improve the ability of each and every member to handle himself in a well-rounded manner during every expedition.

ORGANIZING A CLUB

In most areas you will more than likely run into a fair amount of difficulty in putting together a well-rounded group immediately. These things generally take time, and caving clubs are noted for starting out with three or four members and then expanding as their activities become better known. Publicity plays a great role in gaining attention within the community. Too many clubs start out as a loose organization of friends and don't grow very much from this point, mainly due to a lack of formality in organization and to the absence of any public information program which tells what the group does.

In the beginning it will be necessary to form a group of at least three individuals (including yourself) who know a bit about cave exploring. It may be necessary to continue in this vein for a few months in order to build up needed expertise. When you feel qualified to talk intelligently about caves and cave exploring (after many successful expeditions), it is time to begin to publicize your organization. It is desirable, if not mandatory, for your group to have a name. This should be something that is simple and self-explanatory, such as The Elm City Caving Club. Don't opt for complicated names like The Elm City Speleological and Subterranean Fauna Study Society. This tells the general public little (if anything) about your organization.

Once a suitable name has been chosen, you should plan for an evening meeting to which the public is invited. Proper planning will require you to have an organized program complete with visual displays. The latter can best be accomplished by bringing along a hard hat, carbide lamp, small section of rope, and some of the other gear commonly associated with this sport. Additionally, a slide program is a nice touch and can be quickly put together from photographic essays of some of your previous expeditions. Don't show anything here that might appear to be terribly dangerous to the uninitiated. For example, you may be at a level of expertise where you can descend a 100-foot drop in complete safety, but this might scare the daylights out of someone who has never been in a cave and knows nothing about proper rope work, safety lines, etc. Your slides should include shots of underground streams, cave fauna, cave entrances, explorers, and so forth.

I once put together an introductory program which consisted of a slide presentation taken during an actual expedition. The photographic sequence began with the expedition members loading the equipment in an automobile and included roadside shots along the way and pictures of the pre-entrance preparations at the mouth of the cave. From here, the slide program took the viewer through most of the steps involved in exploring this particular cave. It communicated to the audience the basics of what was involved in exploring the cave and highlighted the fun aspects while still showing the amount of preparation and attention to safety that is necessary. Throughout the program I narrated each event and would pause to answer any questions which arose.

You may be thinking that this is more formally handled than need be, but nothing could be further from the truth. It is necessary to establish an air of seriousness and professionalism in order to attract new members to your organization. Those attending must have confidence in the originators and an intense desire to learn from them. If your meeting is loosely thrown together, then you probably will not communicate effectively with those in attendance. Again, formality is essential, because you want to attract serious individuals who will understand that cave exploring requires strict attention to details and a high degree of obedience and teamwork. If you can't even throw together a simple meeting, then your organizational and leadership abilities during an imagined expedition might be highly suspect.

All in the room must be invited to ask questions, and no questions should be considered laughable. Those who know little or nothing about caves and cave exploring tend to be quite frightened of this sport. At the same time, however, they may also be attracted to the underground beauty and at least willing to consider taking up the sport. During this meeting it will be necessary to answer questions honestly and in a manner which will dispel *unnecessary* fears and myths. Do not sensationalize caving, acting upon the suspected dangers and fears of those less knowledgeable.

For instance, don't say, "In our third expedition, we became hopelessly lost 100 feet underground and didn't know for a while if we'd ever get out." This makes your group sound very amateurish. Should someone ask if you've ever been lost in a cave, you might say, "During our third expedition, we became turned around for nearly fifteen minutes until we were able to find a familiar landmark." This, of course, assumes that you weren't hopelessly lost for hours underground. If you were—and this had occurred recently—then you probably don't have any business trying to tell others how to become a proper spelunker.

In most cases, your meeting should be held in the evening and refreshments should be provided. Choose a meeting place that is adequate to put on a proper demonstration and which will allow typical meeting seating arrangements. Depending on the expected turnout, you might use a public meeting place or even a basement or rec room in a member's home. Stay away from casual "sit-downs" in comfortable living rooms where it is difficult to maintain formality. Friendly discussion groups can be conducted later after a list of potential members has been drawn up.

It will most likely be necessary to advertise the meeting date and time through local media. Most local radio stations and some newspapers offer free announcements of this sort in community event programs or sections. The press should be invited to attend. Nothing helps the formation of a cave exploring club more than a feature article in a local newspaper or an interview on the radio. I would suggest that all speakers during the program be neatly attired. A coat and tie may not be necessary, but this can add greatly to the formality which you are after.

After the presentation has been made and all questions answered, invite the participants to partake of your refreshments. This will give them the opportunity to ask further questions of the speakers and to talk about the prospects of joining your organization among themselves. After a half hour or so, assemble the meeting in formal session again and distribute application forms for interested parties to fill out. These forms can be done on any typewriter and should include basic information such as name, address, phone number, etc. Additionally, a few questions should appear here regarding the applicant's profession, hobbies, military experience, or any other areas which might allow you to assess what the individual might have to offer the group.

A day or so after the meeting, assemble the originators of your group and go over each application form in detail. You might think that anybody who wanted to be a member should be included, but this may not necessarily be so. There are some individuals who love to join organizations and then do absolutely nothing from that point on. You may be able to identify some of these by asking for a listing of other club affiliations on your questionnaire. Undoubtedly you will know some or even most of those who attend your meeting and will be able to make a decision based upon this knowledge. If you don't have this information, however, seek it out before any commitments are made.

Some of the interested applicants may have previous caving experience. Others may have none. At this point, a decision must be made as to what your club is to pursue and to offer each member. For example, if you can select from a fairly large number of applicants with past caving experience, then you might tailor your club to concentrate on explorations with moderate to high difficulty factors. Here, you would not want to include (possibly) an equally large number of people with no experience at all. Certainly, a few inexperienced members might be included at this early stage to allow for continued growth and expansion. If you end up with two equal large groups—one with a great deal of experience and one with none—then problems can develop. You will not be able to satisfy the requirements of both due to the vast difference in experience levels. A few "green" members can be safely included in some expeditions, which can serve as training exercises for them. Eventually, the experience levels will even out. When the two equally sized groups must be provided for, however, this often serves to maintain and even increase the distance between proficiencies.

Chances are, however, that most applicants will have little or no cave exploring experience. Here, the few experienced members of the club most opt for an intensive training program which will involve many hours of "classroom schooling" combined with the eventual exploration of caves with low difficulty factors. This will more than likely try the patience of the few experienced members, but the club will eventually evolve into one which has grown together in proficiency.

In any event, more research is required before you can make a final selection of members. You will, however, be able to weed out a few individuals who it can be ascertained have no business in your caving club. After this initial weeding-out process has been accomplished, you might consider taking all members on a group expedition through a commercial cavern system. In most cases this will be a guided tour led by cavern staff, but you can probably arrange ahead of time to be given a special tour since all members of the group will also be potential members of the cave exploring club. Here, it will be necessary to make arrangements well in advance with the commercial caverns, explaining what your organization is all about and what you hope to accomplish on this guided tour. A special tour of this type will allow members to ask questions about the cave which would probably not be asked by the normal sampling of visitors. If you don't arrange for a special guided tour and simply join one of the standard tours, you will be somewhat limited in what you can ask and what you might be allowed to see. The commercial caverns I have worked for welcomed private tours composed of cave exploring club members. In these instances, special group rates were often offered, and a highly experienced guide was chosen to lead the group. Usually the guide was one who had a good amount of experience in exploring wild caves and was therefore well-equipped to answer some of the more technical questions that might come up. Additional time was allotted for this tour, which might run far longer than a normal visitors' tour. Occasionally tour members were allowed to view portions of the caverns which were normally off-limits.

As one of the organizers of the caving club, your job will be to observe each potential member during the tour. In this controlled underground environment you may easily be able to spot signs of claustrophobia, disobedience, and tendencies toward horseplay. This is not to say that no one should have any fun; this is what cave exploring is all about. This guided tour will certainly not be deadly serious all the way through, but you will probably be able to detect signs of potential problems in some prospective members.

At this point you should be able to make a tentative selection of persons who you think will fit in with and add to your present organization. These should be notified and asked to meet with you and other tentative members to discuss their membership and plan for an actual outing. This can best be held in a nonformal environment, although it will be necessary for you or another member to run the meeting. During the meeting you should outline what is expected in general of every club member. Additionally, each tentative member should be told why he was chosen and how he might serve the organization in a special way owing to his profession or experience. You should also explain that final membership has not been decided upon yet, and will depend upon future participation and performance.

Assuming that you have included a number of people with no caving experience in this group, you should now explain in detail the equipment that will be needed for a first expedition. Each person should be given the opportunity to fill and operate a carbide lamp, wear a hard hat, etc. You will receive many other questions at this point, and each one of them should be answered as simply and as completely as possible. You might then outline the first expedition, which should be within a very easy cave. If you have pictures of the cave, by all means show them. You should also outline in detail any trouble spots that may be encountered, the time of the expedition, and all other pertinent facts. Again, you will be inundated with questions; most of them should be answered, but leave out a few details to allow a certain amount of surprise during the expedition. This will allow you to further rate the performance of these tentative members during the actual exploration.

Before adjourning, all should be advised as to where to purchase their carbide lamps and hard hats. Do not require them to purchase a full exploring outfit, including ropes, ladders, and all the other paraphernalia they may need should they become official members. Make the gear requirements as minimal as safety permits, because some of these individuals may not be selected for permanent membership. A hard hat and carbide lamp will certainly be necessary, along with a flashlight and candles.

Choose a cave for this first exploration that will not require a tremendously long stay. This may allow members to share canteens and other necessary equipment. A simple cave should not require any vast lengths of rope, and certainly should be free of any potentially dangerous drops.

I am assuming here that the total expedition will consist of no more than ten members and that the cave is large enough to accommodate a group of this size. If more members are involved or the cave is rather small, it will be necessary to take only a portion through at one time. This will not require two different days for exploration; one group can remain at the entrance while the other is inside. At no time should any individual be without the immediate guidance and aid of an experienced member.

Once all tentative members have obtained their equipment, it will be necessary to have a dress rehearsal meeting. Here, all persons will be checked out in the operation of their gear. It will be your responsibility to see that all carbide lamps are functioning properly and especially that the carbide container seals are tight. Since lamps from different manufacturers may be used, make certain that any differences in operating instructions are fully understood by the owners. During this get-together, preach cave safety. Let the members know exactly what you expect of them and especially what you expect them *not* to do. Discipline must be laid down strongly during this session, and each individual should know what will happen if orders are not followed. Admittedly, it may be a bit difficult for you and other experienced members to think of an expedition in a simple cave which you've probably been in many, many times as an adventurous outing requiring planning down to the minute details. Remember, you and the other experienced members probably planned your first exploration here in minute detail and know this routine by heart. The inexperienced tentative members, however, have not had the benefit of your experience, so they must be taught from day one what type of planning must be done before any exploration. This means that you must treat the exploration of this simple cave in the same manner you would handle one with a higher difficulty factor.

The day of the first trial expedition will be one in which a final decision can probably be made as to who deserves membership in the club and who does not. Since all members will most likely be meeting at a single location to begin the trip to the cave opening, make note of who is on time. Each member must be thoroughly checked out to make certain that the proper gear has been brought along and that he or she is ready to begin exploring. The trip to the cave opening will be filled with many more questions. Be patient here, as excitement will be running very high.

At the cave opening another tutorial session begins while each member is again checking his gear. Explain again the various high points of the expedition, repeat some of your safety speech, arrange a logical order of progression into the cave, etc. Make certain that you have an experienced explorer at the front of the expedition, one in the center, and a third at the rear. All carbide lamps should be struck before entering the cave.

During each stage of the expedition, note each member's performance and especially his/her ability to follow instructions. Explain to the group as you go along the formations that you encounter and some of what lies ahead. Allow each member to do as much as he/she can on their own, but don't allow anyone to leave the main party without an experienced guide along. All other experienced members of the expedition should also watch each individual and make mental notes as to performance.

Upon exiting the cave, each member should clean and stow his own gear and then everyone should sit down and discuss what has occurred. This should be done before the return trip home and while everything is still fresh and clear in each person's mind. Faults should be pointed out, but an equal amount of praise should offset this. Ask for member reactions; these will tell you a lot about the individual. During the trip home still more information will be relayed. By now it should be fairly clear in your mind as to which members will be offered a permanent place in your organization.

The training program will continue from here, but it will now be possible for all members to purchase the gear needed for future expeditions. The next exploration will include much of the same pre-planning and observance as the first. You will probably notice that some members adapt more readily than others, so it will be necessary to try as much as possible to see that the entire group is moving in one direction regarding experience, proficiency, etc. Newer members will often offer up new ideas, some of which are workable and some not. Some may be tried with modifications. This is where the benevolent results of a good caving club benefit each member, experienced or not. As the group becomes more cohesive, everyone begins to know who to turn to for advice in certain areas. Each person begins to work with the others, and more and more difficult explorations may be attempted.

However, as experience levels increase, caution is still necessary because some of the newer members may tend to become overconfident. From time to time it will become necessary to pull these individuals back in line. For this reason, it is always mandatory that someone be in charge of an expedition and/or the entire club. After a time, some of the newer members may be allowed to lead or co-lead an expedition. This is a somewhat controlled experience, as an experienced leader is always there to lend support should it become necessary.

After this first group of new members has reached an experienced status, you may wish to further expand your organization by repeating the process outlined in these pages. It will be necessary to limit membership in most instances based upon practicality. It is very difficult for large groups (of 20 or more) to successfully explore many wild caves at one time.

While the selection process outlined here may seem time-consuming and intricate, remember that the success of any organization is dependent upon each and every member. One does not go about exploring caves in a haphazard manner, nor does membership selection take place in the same vein. Nothing can spoil an expedition quicker than a member who is uncooperative or needs constant watching. Loosely thrown together organizations tend to collapse rapidly, so the membership selection must be handled with the same care and planning that is mandatory before any underground exploration.

Once you have succeeded in organizing a well-rounded group of individuals to get together on weekends to explore caves, the fun can really begin. A group of ten individuals can often accomplish far more than a small party of three or four. While teamwork is of great importance, individuality is what makes a cave exploring club exciting and rewarding. Some members may wish to pursue cave photography, for instance, and can form a pictorial record of each exploration. You will want to continually keep your organization's activities in front of the public, and these photographs will be quite useful in newspaper stories or possibly even magazine articles. This can also be a means of raising money for club activities, as many publications will pay for the photographs they decide to print.

MONEY MATTERS

It is necessary to discuss monetary matters when dealing with any organized group. Throughout this chapter formality has been stressed, and this will need to be a continuing process. For this reason, the club should meet at least once a month to discuss future plans and expeditions. In most instances, each member will pay his/her own way. This will involve the purchase of equipment for personal use, trip expenses, overnight lodging, etc. Trip expenses are usually shared when explorations some distance from home are undertaken. It is not unusual for cave exploring clubs to travel to adjoining states to explore the more interesting caves, so expense can become a big factor.

The ability of a cave exploring club to raise money from outside the immediate organization will vary, depending upon the location and the actual membership makeup. Some clubs raise money by lecturing on cave exploring; others may sponsor raffles and other events the community can participate in for a small fee. Since it may be necessary to raise money from outside the organization, this is all the more reason for keeping your name in front of the public by using the local media.

It is a good idea to make contacts with local law enforcement agencies and rescue squads to allow them to understand more about your club. Should an accident occur in a cave, the fact that your organization knows about cave exploring and has possibly even explored the particular cave in question can be of great benefit in aiding rescue squad personnel. Occasionally, caves are used as depositories for stolen goods. Here is where your organization might be able to aid a local police department or other law enforcement agency. Whenever one of these services needs your help, every effort should be made to assemble your group as quickly as possible and render as much help as needed. Afterward, make certain that your services are included in any news story which might result. Usually the rescue squad personnel or law enforcement agency will handle this part of your public relations for you. It is also a good idea, however, to make personal contacts with the members of the news media in your area. An even better idea would be to recruit one or more members from media personnel. If you've got your own built-in press agent along on every expedition, you are bound to get more public exposure.

EXPANSION

Assuming that you are successful in getting your name before the public and keeping it there, you may find yourself inundated with requests for membership. While a membership selection process has already been outlined, there may come a time when you simply cannot allow a group to expand further. As mentioned previously, there is a practical limit to the number of people who can be included in a cohesive cave exploring group.

Instead of adding more and more members to an organization that is already large enough, it's a much better idea to ask a few of the experienced members if they might be interested in helping to start up another cave exploring club, one that could work independently but also in conjunction with your organization. By initially forming a cave exploring club and keeping its activities in front of the public, you have inadvertently taken on a great deal of public responsibility. Many individuals in your community may suddenly want to take up cave exploring, and if they can't get expert guidance, some may forge ahead and put themselves in dangerous situations. For this reason, I feel you owe it to those individuals to do everything you can to see that they get the proper guidance. One or more of your experienced members can help form another club, possibly one made up mostly of inexperienced members. This leader might still be a member of the original organization. This should keep everyone happy. Both clubs might wish to work together in fund-raising activities, and the mutual exchange of knowledge between the two organizations should be of great benefit.

At this point it may sound like everybody in the world wants to explore caves and become a part of a caving club. This is not true, and you will probably find that only a handful of people will really want to get actively involved in the sport. In most situations you will not run into the problem of having more applications for membership than you have membership slots available. In rare instances, however, this situation can arise, so it can't hurt to prepare for this possibility.

It is a good idea to consult the National Speleological Society for further information on caves and caving organizations. You will probably want to list your organization with them so that other cave clubs throughout the United States will know of your existence. Sooner or later your group will want to explore a cave in a distant state, and a contact with a local caving club in this area will be of great benefit in your pre-exploration research efforts. The members of this other cave club should be quite familiar with most of their local caves and can be a great help in pointing out those that might be of interest. They can also provide information as to the layout of these caves and the equipment you will need to successfully explore them. Information on landowners, permissions, local weather conditions, and so on will also be available. Every caving club I have ever contacted has always been intensively helpful and courteous in these matters. After all, they may want to explore some caves in *your* area at a future date and may request the same information from you.

Another advantage of an organized caving club lies in the fact that you will be recognized in the community as being more than a loosely formed group of individuals. This can lead to opportunities which would not be available otherwise. One of the most exciting explorations I have ever participated in involved a commercial cavern. The owners were thinking of opening up another section of passageways to their visitors, and my cave club was asked to participate in the full exploration of these passageways and was actually paid a small fee for services rendered. We got to see firsthand how a commercial cavern is prepared for visitors and oversaw much of the work that was done by laborers with no previous experience in caves. We were there to help assure the safety of every worker.

Assuming that all members of the organization have reached a moderate to high level of caving proficiency, difficult caves may be explored in safety and with much more convenience. Let's imagine that your group wishes to explore a difficult cave in an adjoining state. First you would contact a local cave club, if one is known. If not, it might be necessary for two or more members to go on a scouting expedition to find out as much as possible about the cave, permissions, etc., from local residents. These members can then report back to the entire club at a planned meeting, and the entire matter would be discussed in detail. This same scouting party would also have checked on campsites, motel rooms and other practical details which might be involved in a planned exploration that might require spending one or more nights near the cave entrance. Spelunkers often pitch tents near cave openings, setting up an entire camp to serve as the home base through the expedition. Permissions to explore the cave(s), to camp on private property, etc., must all be obtained in advance. A larger caving club is better equipped to handle these details than is a small organization, so the entire expedition, from leaving home to returning again, can be carried out more smoothly. One member might be in charge of medical supplies, another would handle food preparations, still another would be in charge of general equipment. This process would be repeated again and again, with each member responsible for a specific area in the overall expedition. If all these duties had to be carried out by only three persons, the individual workload might be tremendous, and there would be ample opportunity for errors to crop up.

During every expedition and at each meeting, someone should be chosen as club secretary to take minutes and keep a running diary of what occurred. Normally a secretary is chosen to handle all of these assignments for a specific period of time, but sometimes this duty is rotated among the available members. A running diary is most important, because it allows the activities of the club to be assessed at the end of each caving season. This also makes excellent reference data for newspaper articles, lectures, and club reports. Should interest begin to slack off among members, the cause might be more easily identified by referring to this diary and immediate corrections made.

WINTER ACTIVITIES

When the cold winter months arrive, cave exploring sometimes comes to an abrupt halt. Of course, cave conditions remain basically the same in winter as in summer, but braving the cold weather to get to the cave is a problem in itself. Cold weather campouts require much more planning than those in milder weather, and here again a large organization of individuals is better able to cope in a practical manner.

Even during the months when cave exploring is not to be attempted, meetings should still be held every month. These can be devoted to training. Approved methods can be rehashed and rusty techniques brushed up. Climbing techniques may still be practiced using aboveground rock ledges, trees, or even large auditoriums and gymnasiums. The caving club is still a tight-knit group even when aboveground. If members become accustomed to doing things as a group, it will be far easier to get a good attendance at some of these training sessions which may offer far less excitement than an actual underground exploration.

At some of these meetings you will certainly want to devote some time to repairing equipment that has been used for previous explorations. When ropes, carbide lamps, and other caving gear are not to be used for a number of months, it must be put in top-notch working order and then properly stored. Each member can bring along his own personal gear and the entire group can begin repairs and preventive maintenance. Also, a list can be drawn up specifying equipment that should be replaced before the coming exploration season as well as items which the club hopes to purchase to be better equipped.

The winter months are also great times to think about recruiting new members (if desired) and to prepare and practice the presentations that must be given. Fund-raising activities can also be handled during this time. A cave club I belonged to, raised enough money during the winter months to purchase a war surplus vehicle which was used to transport members during expeditions.

Most importantly, a complete activities schedule may be mapped out for the coming caving season. Each member may gather information on a distant cave and present his ideas for exploring it to the club. This need not be done hurriedly, but if everyone does his part, a full schedule of explorations can be ready at the start of the warm weather seasons. A schedule prepared well in advance will allow more members to arrange to attend. The club can also be advised in advance by those who find they cannot attend a specific exploration. When a member is absent, this requires reassignment of responsibilities and possible modification. Naturally, these are best made far in advance of the exploration date.

This is not to say that each and every exploration contemplated during the caving season *must* be put on a planned schedule. Typically, major expeditions will be planned months in advance, but minor explorations can take place almost on the spur of the moment. While a successful caving club depends quite a bit on organization, take care not to restrict members too greatly. If someone suddenly comes up with an idea for an interesting exploration, the group should be flexible enough to be willing to change plans and simply proceed. I'm not getting away from the mandatory planning which is necessary to enter caves, but this does not mean that you must have a step-by-step map of every one you enter. The basic gear discussed earlier in this book should be adequate to allow anyone to safely explore 90% of all caves attempted. The only preparation needed here involves checking with the landowner to get permission and a few other details also discussed previously. Should you enter a cave that is relatively unknown to you and find that you don't have all the equipment this particular one requires, the only thing to do is halt the expedition and try it again later when you are adequately equipped.

Finally, the winter months are an excellent time to hold an annual dinner or banquet to honor members and to invite those who have assisted you during the previous year. Spouses, of course, will be included, along with members of the press, rescue squad, police department, etc. This can go a long way toward maintaining your good public image and continue interest in your club. Some of the law enforcement personnel may be able to comment on any services your club has rendered in the past. In some cases these organizations may even present the club with a service award. This makes for excellent publicity. At this banquet, members can be given club awards for proficiency, improvement, etc. The main point, however, is that this is a time for good fellowship and renewing convictions to continue underground explorations.

Certainly, some readers will say that all of this

preparation and formality is unnecessary, but I would be willing to bet anyone that they can't think of more than one or two long-term organizations that have survived without these types of activities and planning. Loosely-formed semi-organizations of "good ol' boys" who get together every once in awhile and explore some caves just don't seem to last in most instances. The reason for this lies partially in the fact that no one member is really charged with a continuing responsibility. In short, there is no real organization and no rules of discipline. In such a loose organization, each person has his own rules, any of which might conflict with or fail to address those of others. When everyone must participate on an equal basis and abide by the same rules, organization is maintained. This does not severely handicap the individual. All individuals are urged to personally excel to any point they can achieve while still abiding by the rules that apply to everyone.

Sure, the idea of a formal banquet put on by a caving club composed of members who are more often than not covered with mud from head to foot may seem a bit ridiculous. The point here is that you're not covered with mud during the banquet. You'll most likely be dressed in a coat and tie like everyone else. While you may look grubby during and after an exploration, when the mud comes off you will appear very similar to most other human beings. Of course, what you plan in the way of a thank-you dinner will depend upon the number of members in your club and the community in which most members reside. The point here is to continually do something which will allow your community to know of your activities, your organization, and your intent to continue in this pursuit on a professional level.

In summation, it can be accurately stated that there are many, many benefits to forming a caving club composed of ten or more members. Some of these have been discussed in this chapter, but there are many more which will become apparent as organization and cohesiveness continue to develop. Some of the problems have also been pointed out here, and many more will crop up. You should, however, be able to anticipate many of them and apply corrective procedures. Clubs are composed of human beings, no two of whom are alike, so no text can accurately describe all of the problems that you might encounter. The onus is on you and other organizers of such a club to find out as much about each perspective member as possible and then act accordingly and in the best interests of a long-standing organization.

As great as the benefits of a club are, one which is poorly organized from the beginning is worse than none at all. The first membership meeting will probably be the most important step taken; if pre-meeting organization is not adequate, the whole thing could end in catastrophe. A few minor problems at first can sometimes turn into major problems that will haunt you for many months to come.

All in all, however, I think you will find that establishing a caving club will be an exciting and rewarding pursuit. It will add to your enjoyment of underground explorations and will also allow others to participate in this interesting sport in a safe manner.

Chapter 8

Finding Caves in Your Area

It is surprising to those involved in cave exploring to hear comments from non-spelunking friends about the rarity of caves in any given area. When I tell long-time residents of my town that there are over 2500 caves with a ten-mile radius, they are usually quite shocked. Certainly, many areas may be fortunate enough to have a nearby commercial cavern. Such attractions are well-advertised and usually located on well-traveled routes, but most people feel wild caves are few and far between. This is true in some of the northern parts of North America from a practical standpoint. Even here, there are many, many caves. Unfortunately, most of their entrances are buried beneath hundreds of tons of earth and other debris which was brought down by massive glaciers during the Ice Age. Farther south, caves are said to be more numerous simply because the ice layer did not reach this far and the openings are clearly visible.

Even in some of our southern states that contain thousands of caves, the majority of local residents are unaware of their existence. This is due mainly to the fact that caves are sought out by only a very small segment of the population. This small percentage of individuals is made up of geologists, archaeologists, scientists . . . and spelunkers. The latter probably constitutes the largest percentage of this group.

In rural areas the presence of caves may be more generally known. Farmers sometimes find it necessary to fence off cave openings to keep livestock from wandering in. Some cave openings are actually vertical shafts into which an animal or unsuspecting person might fall, so these are often covered or even blasted shut.

While to many people caves are a mere nuisance, others find them interesting and irresistable. Unfortunately, this latter group is often composed of youngsters who seek out the thrill of an underground adventure without the slightest idea of how to prepare for an exploration. For this reason, caves are usually off-limits to them and horrible stories are told by concerned parents in an attempt to assure their obedience. more often than not, these stories add to the adventurous aspects of cave exploring and quite a few youngsters in rural areas have explored more than one cave secretly.

The original horror stories are then repeated when these young spelunkers become parents themselves and try to protect their offspring in the same manner. This is how myths and legends grow. Because of this, a few caves in a given area may be well known, but this may leave hundreds or even thousands of others which remain generally hidden from most of the public.

How do you go about finding caves in a local area? Probably the best method is to go to the site of a known cave and explore in the general vicinity. Wherever you find one cave, you will probably be able to find others nearby. This is not a hard and fast rule, but it is often the case in many areas. If you have a nearby commercial cavern, obtain a map of the general area and then request permission to walk the property involved. You might explain your purposes to the landowner(s) and ask for any information on caves which may be known in the immediate area. The thing to look for when walking the property is outcroppings of limestone. These may appear as giant rock groupings, but look also for any rocky area, regardless of size. While a cave may be found in the side of a hill or mountain with little or no surrounding limestone outcropping, this is fairly rare. Once an area rich in limestone is found, your chances of discovering a cave are far better.

Another way to find cave openings in your area is to talk to some of the old-timers, especially those who grew up in a rural setting. While they may not have explored caves themselves, they will have been around long enough to have heard stories of persons who did and may even know exactly where the cave is located. Any description of the interior of the cave may be useful, but remember that much of this will have been obtained from those who are trying to build up the adventuresome aspects of their exploration. Cave openings that I have heard of from a few old-timers and were described as being 100 feet wide turned out to be more on the order of 15 to 20 feet in width. Nevertheless, a cave opening was in the general vicinity originally described, so this information is worth its weight in gold, although it may not be entirely accurate.

Geological maps can also be of help if available for your area. These may show where concentrations of limestone can be found, these being prime target areas for locating a cave. Once a good site is located, it is necessary to plan a day of hiking and exploring aboveground in an attempt to find a number of exploring possibilities underground. At all times, be sure to obtain landowner permission.

When exploring *for* caves, don't plan on any actual spelunking. This should be a time devoted purely to identifying an area which has a cave or several caves. Once a cave has been found, you may be able to get additional information by asking some local landowners if they've ever talked to anyone who explored it.

Any area which has a cave or two probably has a caving club or at least a number of individuals in the sport of spelunking. A local Chamber of Commerce may be able to provide information here, but a nearby commercial cavern is an even better source of information. Those who work as guides in these attractions sometimes explore wild caves in their spare time, and may be able to tell you about caves in the area or put you in touch with someone who can supply this information.

If no local caving organization or commercial caverns are available, your next best bet is to contact a member of a caving club in another area or state nearby. They may be able to give you some valuable information about caves they have explored or know of in the area where you are searching. The National Speleological Society can supply you with a list of known and approved caving clubs, and is in itself an excellent source of information for caves which may be on their list. This list is prepared partially from information received from caving clubs that have carried out expeditions in various areas.

One method that sometimes works in locating caves is to get situated in an area with known limestone outcroppings and watch for the presence of bats in the early evening hours. Sometimes large groups of bats can be seen exiting even a tiny cave opening and can lead you to a more specific location. This method seems workable in principle but is often quite difficult to put to practical use. It is mentioned here because it has worked in some instances.

Another sign of a cave is a mountain stream which suddenly disappears. The point of disappearance may not be an appropriate opening for human entrance, but it does indicate the presence of a cave which may have an external opening nearby.

In general, some cave openings are quite apparent, while others are very well hidden. The opening to a very large cave may be only a tiny fissure barely large enough to crawl through. On the other hand, a very large opening may lead to a tiny cave. You can't really tell until an actual exploration has taken place. For this reason, do not ignore *any* opening in a limestone outcropping. Any one might lead to a vast underground network. One of the biggest caves in my area is Rodgers Cave (discussed in Chapter 5). The actual entrance is very insignificant and consists of a narrow vertical shaft into the ground. However, once you enter this shaft and descend the 70 feet to the main level, a large network of underground passages and rooms is found.

While all of the techniques for locating caves discussed so far can work at one time or another, probably the most effective way to begin your search involves talking to as many local residents as possible. Many of them will know of or have heard rumors of a cave existing somewhere close by. They may not know exactly where, but sometimes can lead you in a general compass direction to begin a more intensive search. If you like to hike, opt for areas with known limestone outcroppings and keep an eye peeled for cave openings. If you're not specifically looking for a cave, it's quite possible to walk right by one without ever knowing it's there. Remember, many cave openings are very covert. For example, one very large commercial cavern in Pennsylvania went undetected for many years because its massive opening was completely covered by grapevines.

Examine every limestone outcropping very carefully. Camel crickets which inhabit most caves can sometimes be found on these ledges where there is no apparent sign of a cave opening. This can be an indication that the opening may be completely filled in by earth, with a few tiny passages existing that are only large enough to accommodate these tiny animals. Skyline Caverns in Front Royal, Virginia was discovered in this manner. Camel crickets were seen on a limestone ledge, and quite a bit of digging ensued before the explorers broke through to a sizable opening. The presence of camel crickets does not *necessarily* indicate a cave, however. They can also live in small openings that are covered by limestone rock and do not constitute an actual cave.

When I am scouting a promising area for caves, I often carry along a crowbar in order to pry up a few loose boulders in hopes of uncovering a hidden cave entrance. In twenty years of exploring I have found only two such caves, both of which were tiny and not really worth the effort. My lack of pleasing results, however, has not deterred me from continuing this type of search.

Another possible source of information on caves is a local building contractor. Concentrate on those who have built homes in rural areas where you know limestone outcroppings exist. During their site preparations they may have broken through to a cave entrance and been forced to fill it in. If you know that a cave exists in a certain area, there is a reasonably good chance that others will be found in the vicinity. A game warden or forest ranger might also be consulted. These individuals spend much of their time in rural, mountainous areas and also come in contact with hunters, fishermen, and other outdoorsmen who have quite a bit of knowledge about the land. A local extension agent may also be able to provide some assistance.

There are many possible sources for information as to the presence of caves in any given area. Any one may prove fruitful, but chances are it will be necessary to check many or all of these possibilities before local caves are discovered. If the entire search reveals nothing, don't be too disgusted. Consult a geological guide again and begin your own search in earnest. Just because no one you've talked to is aware of a cave in the area does not necessarily indicate that none exist.

Chapter 9

Commercial Caverns of North America

If you have never visited a commercial cavern, you are really missing something. Commercial caverns abound throughout the United States but are more highly concentrated east of the Mississippi, especially through the southern Gulf states. Commercial caverns take in some of the largest and most beautiful caves in North America. These are the caves which were so large and interesting that commercialization became a possibility. Most wild caves will not be nearly as large or filled with formations.

The commercialization of a large cave is a tremendous undertaking. Today, few additions are being made to the list of commercial caverns because of the expenses involved in converting these wild underground wonders. Many of the commercial caverns in the United States were commercialized in the 1930s and '40s; the undertaking was still tremendous, but the expenses were not, due to the availability of cheap labor. Of course, the actual expenses incurred depended upon the changes that had to be made to the natural passageways. Sometimes it was necessary to blast to make the entrance wider. Additionally, some passageways had to be dug out and the walls chipped away to some degree to allow convenient passage of a guided tour. Some of the most beautiful caves had nearly impossible entrances. Here, a type of mining operation might have been required to cut a vertical shaft a hundred or more feet into the ground to intercept an opening room and avoid twisting passageways which required crawling to navigate. Some caves have elevators to take tours into the lower levels and then out again. Others may have wide, level passageways, while some may contain slopes which can be graded for easy access. In every instance, the least expensive conversions are those which require only minor changes in the cave to allow access by non-spelunking visitors.

Of course, every commercial cavern must contain its own lighting system. This is a Herculean task in itself and involves running electrical wiring to every point on the tour. Often the wiring is covered by a thin layer of concrete colored to generally match the limestone walls. Obviously, electrical wiring in a wet environment must meet many rigid standards. Special waterproof cable is required, and light fixtures must also be designed with the moisture in mind. Because there is no natural lighting of any sort in a cave, an extremely powerful lighting system is required. Add to this the necessity of illuminating rooms which are often as big as a gymnasium and the problems of power are multiplied. Lighting systems in commercial caverns are extremely complex, and are naturally vital to business success. Reliability is critical because you obviously can't have a complete failure when several hundred guests may be deep inside. For this reason, many, many different circuits are used. Should one fail in one part of a passageway, another is active only a few yards away.

Walkways are another important part of a successful commercial operation. They are installed after the passageways have been made large enough (if necessary) for tour convenience. Some caverns have installed elaborate concrete walkways, but many try to keep as close as possible to a natural appearance. Here, walkways are made of crushed limestone which may be covered with a thin layer of sand to allow moisture to sink to a bottom layer and thus make walking much easier. The owners of the caverns must decide the tour route in advance and plan it so that it is not too strenuous for young or elderly visitors. Also, provisions are often made to accommodate visitors who are confined to wheelchairs. Steps are commonly used in caverns to allow access from the entrance to the main tour depth level, so accommodating wheelchair tours is sometimes a complex task involving ramps, specialized elevator devices, and possibly alternate routes.

Safety is not only of utmost importance, it is mandated by extremely stiff requirements. Guard rails are commonly seen in commercial caverns and are necessary to prevent accidents. The tour guide is the leader of any visiting group and is responsible for keeping the tour together; warning visitors of outcroppings, low ceilings and other obstacles which can cause a painful head injury; and maintaining the needed discipline. Due to this responsibility, many long hours of training are involved in preparing a guide for his or her first tour. Only a few commercial caverns require visitors to wear hard hats, which are provided just before entering. Most forego this by completing thorough and repeated inspections of the cave ceilings and other possible problem areas. Any questionable formations are removed before the caverns are ever opened to the public.

With all of this preparation, visiting a commercial cave is a treat for the entire family that can be carried out in complete safety. The beautiful natural sights are typically augmented by colored lighting that paints a fairyland scene on the drab limestone rock. While most caverns don't mention this, a bit of showmanship is inherently designed into the cave tour. In other words, a passageway or room that might appear more interesting with the addition of a few more formations is "mechanically restructured." Formations may be removed from one portion of the cave and transplanted in areas along the tour route. Underground streams may be widened and their courses changed for easier viewing. There is certainly nothing dishonest or wrong about this practice, because it is impossible for the tour to visit every portion of the entire cave. When formations are moved and placed along the tour route, it is done as a simulation of how they appeared in their natural setting nearby but in an inaccessible location. Most simulations are made from natural formations, so the fact that they're not in their natural location should make little difference. Although most commercial caverns do not mention the restructuring of formations, if you have a question about a particular formation, the guide can probably inform you as to whether or not it was originally formed at the spot it now occupies or if it has been moved.

It is impossible within the scope of this book to present information on every commercial cavern in North America, but a sampling of caves in various parts of the United States should give you an idea of many of the things to be found on a tour. Most commercial caverns stock brochures on other caves in the same state, or even nationwide. A great deal of information can be garnered from these brochures that will allow you to set up a planned tour of those which appeal to you.

CARLSBAD CAVERNS

One of the most interesting commercial caverns in the entire country is not commercial at all; it is a national park rather than a privately owned enterprise. The massive entrance is shown in Fig. 9-1. Trips in the Carlsbad Caverns are awe-inspiring.

Carlsbad Caverns National Park in New Mexico has been administered by the National Park Service since October 25, 1923. Originally proclaimed a national monument of 720 acres by President Coolidge, its status was changed to a national park by Congress on May 14, 1930. The park now contains 46,755 acres of rugged mountain/desert area on the northeastern slope of the Guadalupe Mountains in New Mexico; it has 70 known caves. Elevations range from 3600 feet to 6350 feet above sea level. A large migratory bat colony occupies a portion of the park's main cavern during the summer. The National Park Service is responsible for planning, constructing, operating, and maintaining all park facilities except for concessions. Management of the park is aimed at preserving its natural features including caves, rock formations, plants, animals, and historical and archeological resources. At the same time, it provides facilities and services to promote safe and enjoyable use of the park by visitors.

Fig. 9-1. Entrance to Carlsbad Caverns. (courtesy Carlsbad Caverns)

Fig. 9-2. The Temple of the Sun. (courtesy Carlsbad Caverns)

Visitation is measured by actual entry into the park as determined through a combination of physical head counts and traffic counter reading. Annual visitation in 1981 was 771,766, an increase of 15 percent over 1980. Since its establishment in 1923, over 24 million people have visited the caverns. Peak travel months are June, July, and August, which account for about half the annual total. Over the past ten years, visitation has grown an average of 4.25 percent per year. The record travel year was 1976 with 876,187; the largest day on record was July 3, 1977, with 12,896.

The park has one developed area at the main cavern. A seven-mile paved entrance road terminates at the visitor center near the cavern's natural entrance. Adjacent to the visitor center are a restaurant, gift shop, nursery, and kennel.

Three miles of the 20.6 mile long main cavern are open to the public. Two options are offered: You can walk into the cavern through the natural entrance on paved trails and tour the entire three miles, including an 830-foot descent and an 80-foot climb on steep ramps. Or you can enter by elevator and tour 1¼ miles of the scenic and fairly level Big Room called the Temple of the Sun. Hand-held radio receivers are provided to each visitor and give interpretive explanations along both tour routes. A scenic part of the Big Room is accessible to wheelchairs. The cavern is professionally lit with over 800 fixtures. Steep trail sections have handrails. Near the halfway point on the complete tour and at the lower elevator terminus where the Big Room tour starts are a concessioner-operated lunchroom and restroom facilities. Four high-speed elevators bring visitors back to the visitor center on the surface at the end of both tours. Figure 9-3 shows another section of the Big Room, the Hall of Giants.

There is also an amphitheater at the natural entrance to the cavern just east of the visitor center. This theater is used for evening bat flight programs during the summer as well as other special events. A ten-mile, one-way loop drive over a graded gravel road offers views of Rattlesnake Canyon and Upper Walnut Canyon.

A primitive cave experience is also available at undeveloped New Cave, 23 miles from the main cavern in the Slaughter Canyon area. Reservations and advance information are necessary for this tour, which involves some outdoor hiking and climbing, in addition to the 1¼ mile cave tour by lantern and flashlight.

Picnicking is available at the Rattlesnake Springs unit, eight miles southwest of the park entrance and White's City. Several tables are also available next to the visitor center parking area.

The park entrance fee is $3 per carload daily, or $1 per person for commercial bus passengers. A separate user fee of $2 per person, age 16 or over, for the guided primitive New Cave trip has also been implemented. Annual Golden Eagle and Golden Age permits are honored.

Park Rangers regularly patrol the main cavern trails, entrance road, and other areas to provide information and assistance, as well as to insure protection of park facilities and public safety. Rangers also escort visitors on the New Cave lantern trips and other periodic special cave trips. Bat programs are given at sunset each evening at the cavern entrance amphitheater from May through September.

Trips into Carlsbad Caverns are available every day of the year. Reservations are not necessary. All trips are continuous and self-guided and may be started at any time during regularly scheduled hours.

The complete Walk-In Trip (three miles) begins at the cavern's natural entrance near the visitor center and usually takes 2½ to 3 hours. However, visitors may remain in the cavern as long as they wish, up to the departure time of the last elevator to the surface. In the first 1½ miles of walking, visitors descend to a depth of 830 feet below the surface, then ascend 80 feet before reaching the Underground Lunchroom, where the elevators are also located. The second portion of the trip starts at the Underground Lunchroom and is a relatively level, easy walk of 1¼ miles around the Big Room. Visitors wishing to take only the Big Room Trip, which requires approximately 1½ hours, may enter the cavern by elevator. Everyone returns to the surface by elevator.

Photography, including use of flash and time exposures, is permitted on all trips. However, all photos must be taken from the paved trails, without resting camera equipment on cavern rocks or formations.

Box lunches, individual sandwiches, hot and cold drinks and other items are available in the Underground Lunchroom. Surface concession facilities include a restaurant, gift shop, nursery, and kennel service. Visitors in wheelchairs may enter the cavern by elevator and visit a portion of the Big Room. Trails in other portions of the cavern are either too steep or narrow, or include stairs and cannot accommodate wheelchairs.

The cavern temperature is a constant 56 degrees, so a light jacket or sweater is recommended for the trip. Low-heeled shoes or rubber soles or heels are also recommended. Shoes with heels made of plastic are not safe, as they have a tendency to slip on the steep downhill grades.

During the first 1¾ miles of maintained trail, the visitor descends more than 800 feet, passing many of the cavern's most beautiful formations. Figure 9-4 shows the King's Palace, which is seen on this descent. Figure 9-5 shows the frozen waterfall in the Green Lake Room in another area of the cavern.

NEW CAVE

As mentioned earlier, New Cave is an undeveloped cave. It is located near the mouth of Slaughter Canyon 23 miles from the visitor center of Carlsbad Caverns National Park and 36 miles from

Fig. 9-3. The Hall of Giants. (courtesy Carlsbad Caverns)

Fig. 9-4. The King's Palace. (courtesy Carlsbad Caverns)

Fig. 9-5. The Frozen Waterfall in the Green Lake Room. (courtesy Carlsbad Caverns)

the city of Carlsbad. To participate in this undeveloped cave trip, you must furnish your own transportation to New Cave and be prepared to walk one mile from the parking area at the canyon mouth to the cave, 1¼ miles through the cave, and one mile back to your vehicle for a total of 3¼ miles. The walk to the cave is strenuous over a primitive trail with an elevational rise of 500 feet to the cave entrance and should be attempted only by those in good physical condition. Allow 30-45 minutes for the hike to the cave so you can make the climb at a leisurely pace. It is also highly recommended that visitors stay on the trail at all times, since shortcuts are a safety hazard and also cause erosion.

Tours of New Cave are scheduled daily during the summer. During the winter season, which begins in early September, tours are offered on weekends only. Exact times of trips are available through the visitor center and reservations may be made a day in advance. The National Park Service reserves the right to cancel or alter tours as necessary.

Items necessary for the tour are: one flashlight per person, hiking boots or good walking shoes, and water. The cave temperature is a constant 62° with the humidity in the 90s. Infant carriers and backpacks are not recommended for this tour. Photography is permitted but it is requested that tripods not be used. Figure 9-6 shows a map of the general area of New Cave and its proximity to Carlsbad Caverns and White's City.

LOST WORLD CAVERNS

The Lost World Caverns, located in the

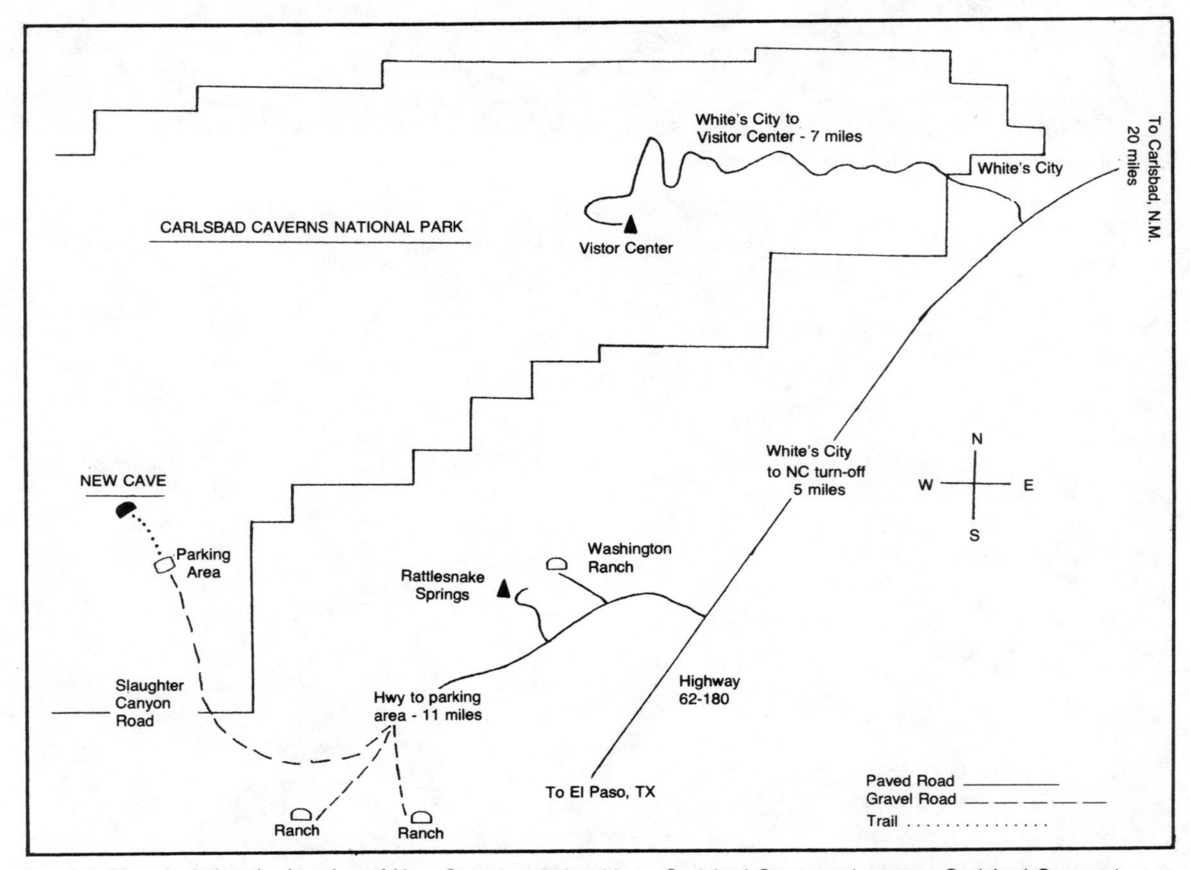

Fig. 9-6. Map depicting the location of New Cave in relationship to Carlsbad Caverns. (courtesy Carlsbad Caverns)

Fig. 9-7. Formations in Lost World Caverns. (courtesy Lost World Caverns)

Greenbrier Valley near Lewisburg, West Virginia, were discovered in 1942 by speleologists using ropes and rope ladders. It was known as Grapevine Cave, being named for the immense wild gravevines that draped over the natural 120-foot pit entrance. In 1946 a group of cave explorers from Charleston, West Virginia, lowered themselves down into the pit, which was the height of a ten-story building, and began the work of making maps, pictures, and exploring the rooms. Serious thought was given to commercial development in 1952, when the man-made tunnel entrance was conceived. Fifteen years later, in 1967, the wonders of Grapevine Cave were made available to the public as Lost World Caverns. The Caverns were closed in 1978 for extensive modifications, and reopened in the spring of 1981 with a new entrance tunnel, new walkways, and a new reception center and gift shop.

Speleothems of one kind or another are found in almost all caves; stalactites, stalagmites, and flowstone are the most common speleothems. Most wild (undeveloped) caves are barren tunnels and passageways or have only scattered and insignificant speleothems. Lost World Caverns is an exception, with its profusion of large spectacular stalactites, stalagmites, draperies, flowstone, and columns. The large size of the rooms, with magnificent panoramic views, are more comparable to some of the larger western caves than its eastern counterparts. A trail encircling the rooms provides a close-up view of the formations that were so highly prized by their discoverers. The visitor walks this trail at his own pace, unhurried, and at his leisure to enjoy the formations that have been viewed by thousands of visitors.

The decorations of Lost World Caverns are among the most spectacular cave formations in the United States, and they occur here in their most beautiful form. Scattered throughout Lost World, in addition to stalactites and stalagmites, are many other types of dripstone and flowstone speleothems, such as soda straws, pencil stalactites, column, draperies, and bacon rind. Figures 9-7 and 9-8 show some of the formations inside Lost World Caverns.

The formations in Lost World Caverns are innumerable, from small to enormous. How old are they? The cave guides usually answer, "One cubic inch in one hundred years;" but of course, that is governed by moisture conditions and volume of air movement. By that rule, Goliath, which is eight feet in diameter and forty feet high, and considering that it developed half as a stalactite and half as a stalagmite, would be a little more than 130 million years old, give or take a million years or so. The hex stones that border a portion of the walls of the main room, having been formed by the floors of the prehistoric oceans, are most interesting. After consideration of these facts, Lost World Caverns has been designated a Registered Natural Landmark by the United States Department of the Interior.

Lost World Caverns is open from March 15th to November 31st, from 9:00 a.m. to dusk. Group appointments are available year-round. Admission is $4.00 for adults and $2.00 for children. The entire walking tour usually takes approximately one hour. There are many lit information signs to make the tour both interesting and educational. The caverns is located in an area with many fast food restaurants and several gourmet restaurants as well. Motels and camping sites are also nearby.

CRYSTAL CAVE

Crystal Cave is located in the heart of the Pennsylvania Dutch Country, midway between Allentown and Reading near Kutztown. It is open every day from Washington's Birthday in February through October 31st. In November, it is open Fridays, Saturdays, and Sundays only, and closes after Thanksgiving weekend.

Crystal Cave is a dazzling arena of crystal magnificence carved slowly by nature through the ages. There is an abundance of milky white stalactite, stalagmite, and dripstone formations which seem to be the columns of a grand palace whose walls and ceilings are adorned by nature. In this cavern you will see delicate flutings and ornamentation in many different tints and shades of color, all enhanced by indirect lighting. The 45-minute tour with a trained guide is along concrete walks with steel railings.

Some of the more outstanding formations in Crystal Cave are the Cathedral Chamber, the Prairee dogs, the Giant's Tooth, the Ear of Corn and Tobacco Leaves, the Natural Bridge, the Indian Head, the Totem Pole, and the Crystal Ballroom. Figure 9-9 shows one of the more impressive columns encountered on a tour through Crystal Cave.

Discovered in 1871, Crystal Cave is the oldest operating cave in Pennsylvania and one of the most heavily patronized national attractions. The caverns were re-illuminated in 1974 and an educational slide presentation, "Inside the Earth," is a part of each cave tour.

Figure 9-10 shows the Dutch Food Center and Hex Souvenir Barn located near the entrance to the cavern. Other things to do and see at no charge include farm animals, playground, nature trail, authentic Amish buggy, real Indian tepees and totem poles, tree plantation, and a picnic park, all on 125 scenic acres. There is also a new, professionally designed miniature golf course, which is three-shaded and cave-oriented. Motels, campgrounds, and accommodations serving Pennsylvania Dutch dinners are all located nearby.

MAMMOTH CAVE

Mammoth Cave is located in the forested ridges of Mammoth Cave National Park in Kentucky and is operated by the United States Department of the Interior's National Park Service. It is advertised as the world's longest cave, exceeding the presently known 226 miles of its length.

The cave's history, spanning 4000 years, is as extensive as its length. The first explorers were prehistoric Indians who searched for minerals on the cave walls. These early people left behind cane

Fig. 9-8. Another view of Lost World Caverns. (courtesy Lost World Caverns)

Fig. 9-9. One of the columns seen in Crystal Cave. (courtesy Crystal Cave)

Fig. 9-10. The Dutch Food Center and Souvenir Barn at Crystal Cave. (courtesy Crystal Cave)

reed torches, moccasins, and gourds as evidence of their visits.

During the War of 1812, Mammoth Cave became a main supplier of saltpeter, which was used to make gunpowder. The tradition of touring the cave began a few years later in 1816. These old-time trips were all by lantern light, with torches being thrown on ledges and side passages. Today, nearly a half-million visitors a year tour this underground system. Recently dedicated as a World Heritage Site, Mammoth Cave is now a place for all to enjoy.

Today's tours through Mammoth Cave proceed through a variety of passages and include many historical and geological features. Of special interest are the prehistoric Indian artifacts, gypsum flowers, vertical pits and domes, and the immense passageways and rooms. The cave temperature is 12°C (54°F), and a jacket is recommended, along with comfortable walking shoes for the strenuous walk.

Mammoth Cave offers a number of different tours of varying lengths. The Historic Tour is a two-mile, two-hour trek featuring a War of 1812 mining operation, a torch-throwing demonstration, Indian artifacts, cave life, Fat Man's Misery and the Mammoth Dome. The mining operation is shown in Fig. 9-11.

The Frozen Niagara Tour features an impressive assortment of cave formations including stalactites, stalagmites, flowstone, and draperies. This 1½ hour tour is the least strenuous tour available, requiring a half-mile walk.

The Half Day Tour features narrow winding

cave passages, dome-pits, breakdowns, gypsum flowers, and cave formations. This strenuous tour requires a four-mile, 4½-hour walk, including 700 steps, steep hills, low passages, etc. Food is available (not included in the tour fee) in the Snowball Dining Room one hour after the tour starts. The Half Day Tour also includes the Frozen Niagara Tour. Figures 9-12 through 9-14 show some close-ups of the delicate gypsum flowers which are encountered during this tour.

Yet another tour is the Lantern Tour. This is a trek by lantern light which features the natural cave entrance, the War of 1812 mining operation, Mummy Ledge, tuberculosis huts, the largest cave rooms, Indian artifacts, and a torch-throwing demonstration. The tour includes a half-mile of the Historic Tour route. The walking distance is about three miles and the tour duration is about three hours.

The Great Onyx Tour features an assortment of cave formations, stalactites, stalagmites, helictites, etc. The tour includes a three-mile bus ride through a hardwood forest with interpretation of the surface features. It covers a walking distance of about one mile, which includes a quarter-mile walk from the bus to the cave entrance. The tour duration is about 2½ hours.

Mammoth Cave National Park offers many other activities, including a number of guided walks on the surface, evening programs, and environmen-

Fig. 9-11. War of 1812 mining operation. (courtesy Mammoth Cave)

Fig. 9-12. A closeup view of gypsum flowers. (courtesy Mammoth Cave)

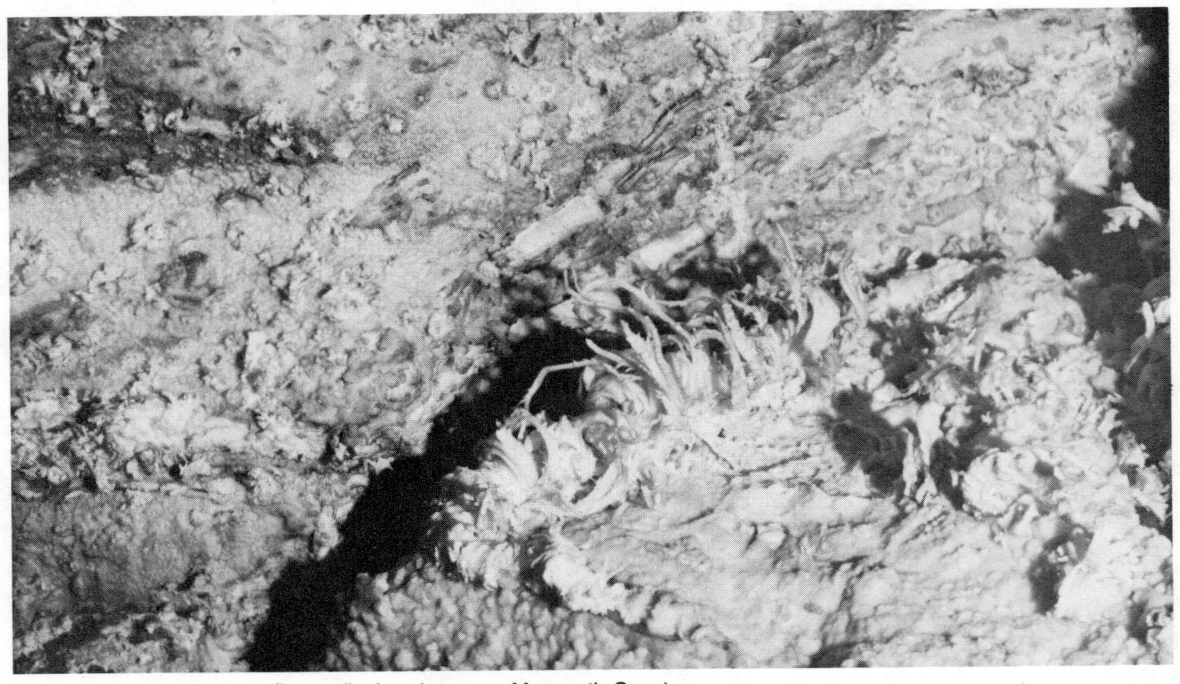

Fig. 9-13. Another gypsum flower display. (courtesy Mammoth Cave)

tal education classes. Further information can be obtained by writing to Mammoth Cave National Park, P.O. Box 68, Mammoth Cave, Kentucky 42259.

CASCADE CAVERNS

Cascade Caverns is located about fourteen miles northwest of San Antonio, Texas, and is in a 105-acre park which includes a campground, picnic grounds, swimming pool, dance pavilion, and a meeting hall for group activities. For many years, inhabitants of the famous and beautiful Hill Country of Texas have known of the spectacular entrance to Cascade Caverns. Artifacts of the ancient Indian tribes that have lived in this region have been found in and around the Caverns. Also, there is a legend that an early German settler lived as a hermit in the first room some 100 years ago. Two books have been written about this man one in 1876 in German called *A Wasted Life*, and one in 1932 in English entitled *The Hermit of The Caverns*.

Cascade Caverns are water-formed caverns on the edge of the Edwards Plateau, just south of the old German town of Boerne. At this point, the plateau is 1400 feet above sea level and 700 feet above the center of San Antonio. The caverns are considered 95% active. Millions of glistening drops of pure water hang from the tips of the glistening stalactites, splash down upon the sparkling stalagmites, or flow silently across the faces of the colorful dripstone formations, which are still in their age-long process of growing. Over 100 million

Fig. 9-14. Here, the flowers can be seen in detail. (courtesy Mammoth Cave)

Fig. 9-15. The Cathedral Dome and view of one body of water. (courtesy Cascade Caverns)

years ago, the ceiling and walls were an ocean bed, and shells and fossils of marine life are plainly visible in many places.

Mammoth molars, Ice-Age fossils, and bones and tusks of prehistoric animals that roamed this region in the distant past of the Pleistocene Period before the age of man have been found here. A trip into the caverns takes you into carved corridors and through room after room decorated with awesome sights of infinite variety. As you gaze at the great waterfall or try to comprehend the mighty forces that carved the rugged Strom Canyon, time goes back millions of years.

A guided tour of Cascade Caverns takes from 45 minutes to one hour and is educational as well as enjoyable for all ages. The walkways are clean and easy to travel, and the passages are well-lit. The caverns are approximately one-third mile long, with a depth of around 190 feet below the surface; average temperature is 65°F.

Cascade Caverns takes its name from the beautiful waterfall which plunges almost a hundred feet from a shallow cave containing an underground stream into the main cave. The caverns are open every day of the year. Figure 9-15 shows a view of the Cathedral Room and one of the bodies of water in this system. Additional information can be obtained by writing to Cascade Caverns, Route 4, Box 4110, Boerne, Texas 78006.

SKYLINE CAVERNS

Skyline Caverns is located in the Blue Ridge Mountains one mile south of Front Royal, Virginia, near the entrance to the famed Skyline Drive. It is the closest major caverns to Washington, Baltimore, and other major population centers of the Northeast.

Estimated to be thirty million years old, Skyline Caverns were discovered in 1937 by the late Dr. Walter S. Amos. They are the only caves open to the public known to have been discovered by scientific deduction. According to Skyline

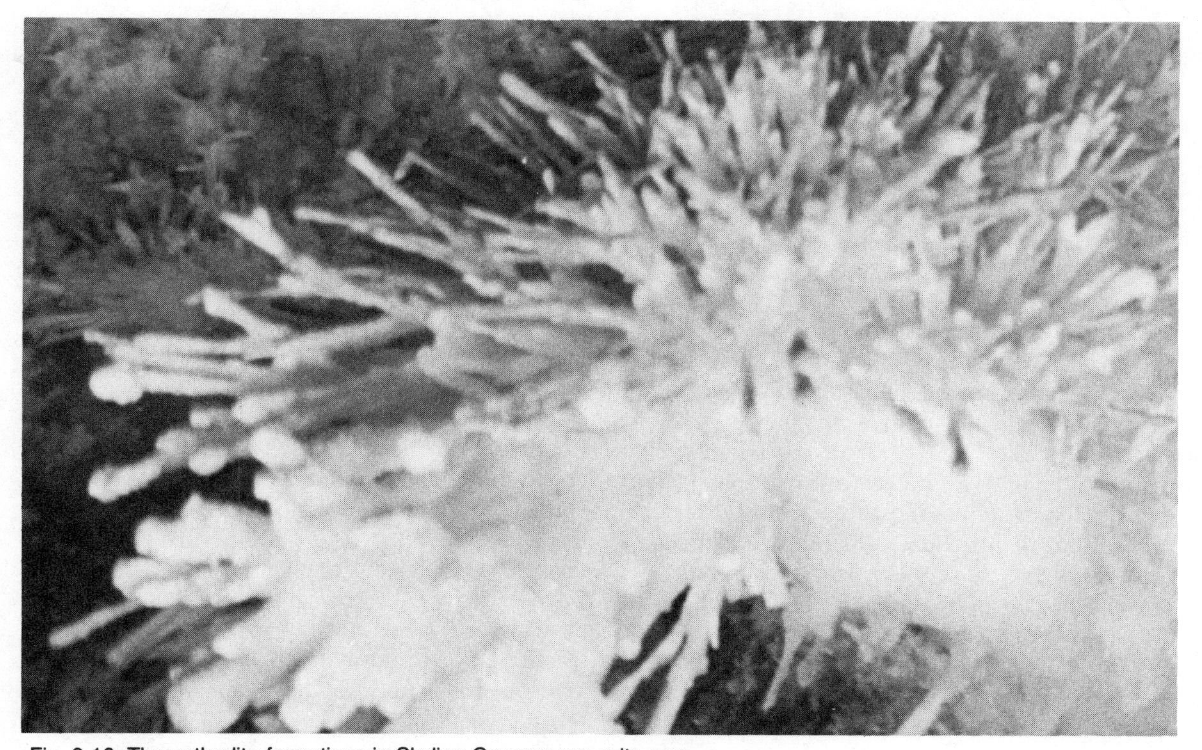

Fig. 9-16. The anthodite formations in Skyline Caverns are quite rare.

Fig. 9-17. The Capitol Dome in Skyline Caverns.

Caverns information, this attraction has a constant temperature of 54° F and widest variety of formations of any known cave in the world. In addition to outstanding examples of stalactites and stalagmites, it contains the unique anthodite formations.

Known as "the orchids of the mineral kingdom," anthodites are delicate formations of pure white calcite and puzzled geologists because they grow out in all directions, apparently in defiance of the laws of gravity. Their growth rate is estimated at one inch every 7000 years. These formations are shown in Fig. 9-16.

Descriptive names have been given to other unique formations, passages, rooms, and bodies of water in the caverns. These include the Capitol Dome, Rainbow Trail, Painted Desert, Fairyland Lake, The Shrine, and The Trout Stream, shown in Figs. 9-17 through 9-22. The Trout Stream is stocked by the Virginia Department of Game and Inland Fisheries as a conservation experiment. Skyline Caverns also contains a magnificent waterfall (Fig. 9-23).

Many findings of scientific interest have been discovered in Skyline Caverns, especially in regard to rare animal life. A cave beetle, born without eyes, is found here and no other place in the world. Many areas of Skyline Caverns are still being explored by scientists and may eventually be open to the public.

Skyline Caverns has been officially recognized as the best-illuminated cavern in the world. All wiring is covered with a compound which blends in with the cave walls. This is an indirect lighting system and visitors never look directly into the glare of an incandescent lamp. Certain areas of the Caverns are so well-lit that it is possible to take photographs without flash bulbs.

Of special interest is a recorded program called "God in the Mountain," which takes place in Cathedral Hall, the largest room on the hour-long tour. Narration and music are synchronized with changing light patterns for a very dramatic effect.

Being a major tourist attraction, Skyline Caverns also offers a restaurant, gift shop, and miniature train ride. Annual events at this attraction include the Skyline Run, a widely attended antique automobile show, and the Hudson Essex Terraplane Meet. Food and lodging are available in nearby Front Royal, Virginia on a year-round basis that matches the caverns' operational year. Skyline Caverns are open every day of the year, except Christmas Day. Each tour takes approximately one hour, and the last scheduled tour on any given day starts 30 minutes before closing time (5:00 p.m. in the winter months; 6:00 p.m. in spring and summer). Caving clubs who might wish to visit Skyline Caverns may make special arrangements with the management, who will be happy to open their attraction prior to normal opening hours or will remain open after hours in the evening. A minimum of ten days' advance notice must be given so that confirmation of special hours can be made.

Fig. 9-18. Rainbow Trail. (courtesy Skyline Caverns)

Fig. 9-19. The Painted Desert.

Fig. 9-20. Fairyland Lake.

Fig. 9-21. The Shrine. (courtesy Skyline Caverns)

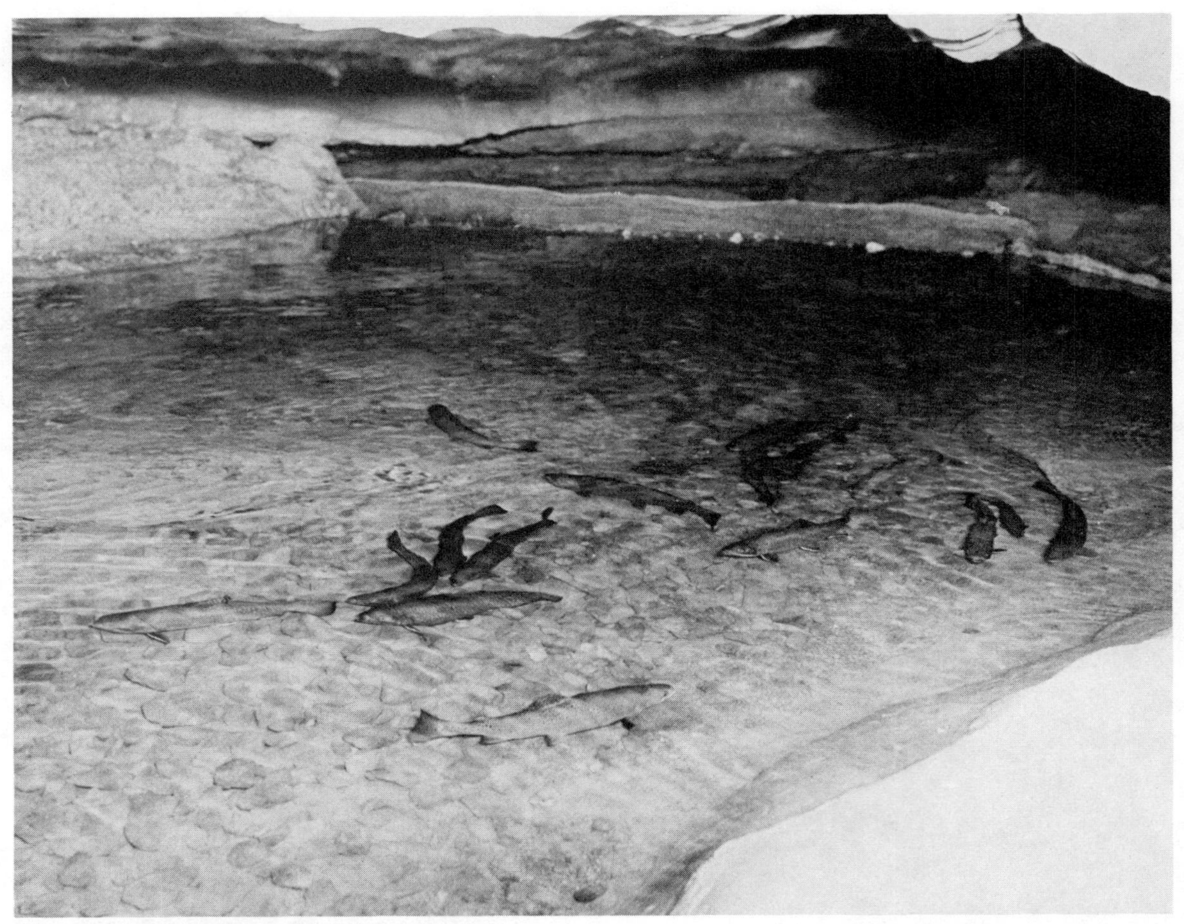

Fig. 9-22. The Trout Stream is stocked regularly.

LURAY CAVERNS

Luray Caverns, located in Luray, Virginia, in the heart of the Shenandoah Valley, has a most fascinating history and features the unique Stalacpipe Organ, a musical instrument designed to produce musical tones by tapping against the stalactites in the caverns. The discovery of Luray Caverns a century ago was probably the worst-kept secret in the eons-old history of the Shenandoah Valley, if not in the entire land. Even after the presence and potential of the subterranean wonder became known to more than a few (only hours after it was found), buyers of the land above it were ridiculed for paying $17.50 an acre for very poor farm land full of rocks.

Andrew Campbell shares the discovery credit with several others, but it was the speleological urge of a migrant photographer from New York State that opened up the Luray legend. Benton Stebbins and his wife, Amelia, were attracted to the Shenandoah Valley by the glowing promotion of the Shenandoah Valley Railroad which was then under construction. After a five-day wagon trip from Easton, Maryland, Stebbins pitched his office tent on a vacant lot in Luray, put an ad in the *Page-Courier* and opened the Stebbins Photo Studio on June 27, 1878. He immediately began to reason that any country with such an abundance of limestone should have caves.

In 26-year-old Billy Campbell, Stebbins found

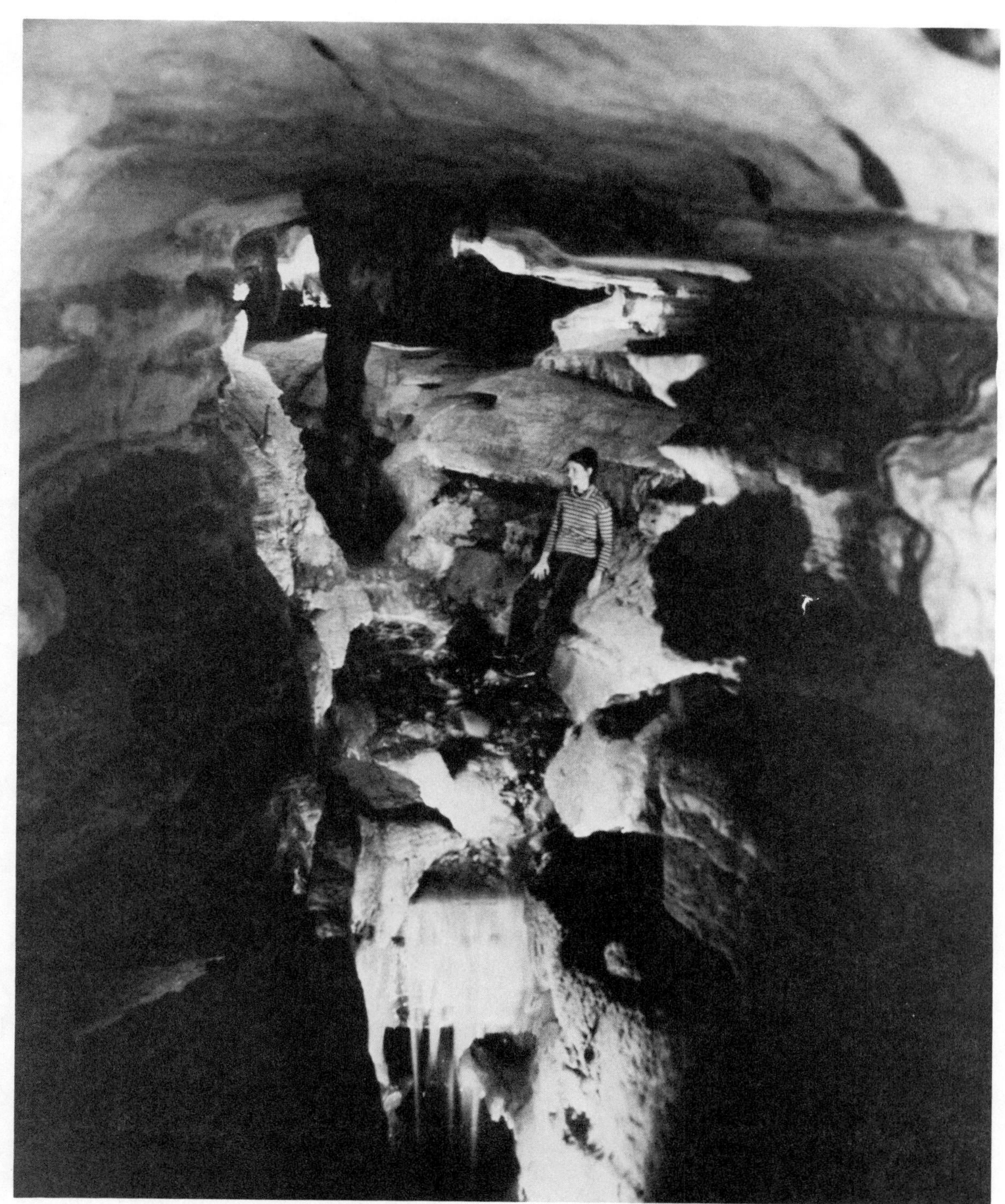

Fig. 9-23. This waterfall presents a magnificent sight.

not only a fellow photo fan, but a man more than slightly interested in the underground. Campbell's Uncle Andrew had taken him into a nice cave known as Ruffners. The caverns are under the same hill and adjacent to today's Luray system.

On the morning of August 13, 1878, an exploring party led by Andrew Campbell, Stebbins, Billy Campbell, and accompanied by James Modisett, Andrew's cousin, and 13-year-old Quint Campbell, found a hole emitting cold air. They shoved aside a few rocks and discovered Luray Caverns. Andrew Campbell went down a rope and out of sight, closely followed by Quint. Candlelight opened up an unbelievable preview of what was there in seemingly unending splendor. Still below ground, Andrew and Quint took a secrecy vow; upon surfacing, Andrew assayed it as "nothing but a hole in the ground," thereby setting the groundwork for exploitation. No sooner was James Modisett out of earshot, than Andrew Campbell told Stebbins what he so wanted to hear. Then Billy Campbell joined the secrecy pact.

In less than 24 hours, there was a meeting to explore purchase of the 28.5 acre site owned by S.A. Buracker. An immediate problem was raising the estimated $600 for the purchase. They didn't have that kind of money among them, but they knew Buracker would be willing to sell. He had been a wealthy Luray merchant before the Civil War, but sickness and non-paying customers had left him on the brink of indigency. Much of his property went up for auction, but the cavern site was not included among the bids.

In the meantime, a leak developed. Billy Campbell, attempting to raise his share of the purchase price, appealed to his father, the Sheriff of Page County, who automatically and financially became a partner. Finally, on September 19, 1878, the Campbell-Stebbins combine submitted a winning bid of $17.50 an acre, a figure considered exorbitant at the time. (Similar land in the cavern area would go for up to $3000 an acre today.)

Further exploration followed, and the news abounded even before the *Page-Courier* announced that land values were increasing in the Cave Hill region. After visitors were being admitted for 50 cents, the newspaper began to tell it all: "A Wonderful Cave . . . Subterranean Vaults of Mammoth Dimensions . . . Columns and Pillars of Stalactites and Marble-Like Whiteness . . . Marvelously Beautiful . . . Incredibly Grand . . . Crystal Springs and Silver Lakes . . ."

Further exploration had its macabre aspects. Andrew Campbell discovered what appeared to be the skeleton of a 12-year-old boy, the first evidence of human presence in the caverns in a far earlier time. The Luray discovery attracted national interest with a story by Major Alexander J. Brand, Jr., in the New York *Herald* on October 19, 1878. Major Brand, after his visit, said, "Trying to compare your cave to others would be like comparing New York City to the Town of Luray."

After an official opening date had been set for November 6, 1878, the *Herald's* publisher, James Gordon Bennett, assigned a staffer to cover the Luray story. Professor Jerome J. Collins, the writer-explorer, postponed a three-year trip to the North Pole despite misgivings. Professor Collins said he found excitement high among residents and that every sinkhole in the country was being investigated with the hope of finding another cave of Luray's proportions. But curiosity and spelunker fever hit new publicity highs when the Baltimorean published a story that the Luray Cave flap was a hoax, stirring a controversy that produced more publicity.

The first illumination in the cavern was produced with a locomotive headlight borrowed from the Chesapeake & Ohio and supplemented by cave browsers' candles. By late December 1878 almost a mile of trails had been developed, and a wooden floor was put down in Giant's Hall in time for the second illumination on December 27th. This was to produce winter revenue for depleted construction funds.

Suddenly, it seems, Benton Stebbins felt the photography urge once more to help sell the cavern story, but couldn't come up with adequate lighting. His wife made India ink drawings which Stebbins photographed. But caverns management failed to interest officials of the soon-to-be completed Valley Railroad. Even the Smithsonian Institution rejected early invitations to help feed the promotion machine, but the word was getting out.

By mid-morning of the day of the second illumination, wagons and carriages blocked the road leading to the cave. Over 600 visitors paid $1 each. Then *Harper's Weekly* published an article and the New York *Herald* gave it another impressive plug.

With all this came a legal cloud. Original landowner Buracker wanted his land and cave back, but after long legalizing, a court held that the original secret pact did not constitute fraud. The cave operators celebrated by hiking cave tours from 50 cents to $1. Tiffany's of New York displayed some stalactites. The New York *Times* finally got the cavern message and the stage line connecting Luray with the railroad in New Market fourteen miles away was doing a flourishing business. Finally, on July 12, 1880, the Smithsonian was convinced there might be a major discovery. A written report was issued stating that "Comparing this great natural curiosity with others, it is safe to say that there is probably no other cave in the world more completely and profusely decorated with stalactite and stalagmite ornamentations than that of Luray."

As the completed Valley Railroad brought its first train into Luray in October of 1880, rail executives took another look and decided they wanted all of the action and made a preliminary offer of $40,000. Even as Stebbins took the offer back to his partners, the maze of legalities pursued by the Buracker interests droned on, based largely on whether existence of the cave was indeed a secret at the time of the original sale. A Richmond court ruled that the tract once more should be offered for sale (cave and all) at public auction with an opening bid of $10,000 which William T. Biedler, a Buracker son-in-law, proposed to make. Biedler, understandably elated by the ruling, already had conversations underway on the purchase of the cave by the railroad. During these conversations he rhapsodized on the beauty, size, and coloring of the cavern, but when confronted by a rail lawyer, admitted he had never been in it. Benton Stebbins, who had by this time become a cave owner, schoolteacher, newspaper publisher, carpenter, medicine show salesman, and Mayor of Luray, realized he had lost.

Biedler sold the land for $39,400 to the Valley Railroad through the Valley Land and Hotel Company, a holding group formed by the railroad. Benton Stebbins and Sheriff Campbell received nothing. Andrew J. Campbell and Billy Campbell remained as cavern guides.

Near the turn of the century, the holding company went bankrupt after a fire destroyed the original Luray Inn and a depression crippled the nation's economy. The enterprise was purchased by Colonel T.C. Northcott, who formed the Luray Caverns Corporation, a family-owned business. Good management and a highly efficient promotion program has since made the caverns known throughout the world and an attraction visited by a half million visitors annually. The Stebbins and Campbells found no rich ores in the caverns, but the system is both fascinating and beautiful in itself.

In 1954, Mr. Leland W. Sprinkle visited Luray Caverns and, as a result of a guide's demonstration, learned that stalactites produce musical tones when gently tapped. An electronic scientist and organist, he became excited by his newly-formed idea of creating an instrument for playing musical selections on the stalactites.

In the years following, he developed The Great Stalacpipe Organ at Luray Caverns, a musical instrument not only unique but also the largest in the world—and still growing. It can be made to include all tunable stalactites in the 64-acre caverns. To "sermons in stone," he has added "music in stone," unlocking the tones imprisoned in silence, darkness, and hardness for countless centuries underground. The caverns were formed by underground rivers and seepage of acid-bearing waters through layers of limestone and clay. Eventually the clay was washed away, leaving the great chambers and avenues of stone that give the caves their intriguing shapes. But for all the millions who have been enchanted by the strange and exquisite formations, it took just the right combination to discover and and develop the secret of their music. It took the curiosity of a five-year old child (Robert Sprinkle, who wanted to see Luray Caverns as a birthday gift), the scientific genius and musical imagination of the father, and the adventurous faith of the staff to allow the exciting experiment.

Starting from scratch, without precedent, Sprinkle had to solve many unusual problems. To begin with, was it possible to tune a stalactite? Sprinkle found only two that were naturally in tune. A system of sanding the others had to be worked out with aluminum oxide discs rotated at high speed. Although the discs are strong enough to wear down tempered steel, the stalactites are so hard that they wear them out quite rapidly.

A set of 13 English tuning forks was used for initial prospecting. Then precise tuning was done by reference to a system of precision tones amplified so that they could be heard above the sanding, which continued until the beats or wavy effects disappeared. Once tuned, the extraordinarily hard rocks maintain their pitch quite well. Air-conditioned by nature, they are maintained at a constant 54° Fahrenheit. It might be a good idea to check the large stalactites, says Sprinkle, at least once every thousand years. By skillful use of the sander, stalactities often have their tone qualities improved by the elimination of unwanted harmonics.

It has taken Sprinkle years of research, design, and experimentation to develop the octave chassis for the Stalacpipe Organ. The overall design of the instrument is such that the driving chassis for each octave is a complete and separate block, with power supply, solid-state circuitry, and firing apparatus. All blocks are interchangeable and spares are maintained for split-second replacement. A block operating the bass, for example, can be interchanged with a block operating the treble without affecting operation of the organ.

When being played by the robot player, the octave blocks are controlled by metal brushes simlar to tiny paint brushes. They rest on a thin but extremely tough and durable Mylar plastic belt, three-thousandths of an inch thick, 40 inches wide and over nine feet long with patterns of slots melted into it with a soldering iron. These slots correspond to various chord structures and other musical configurations derived from a system of numbers representing tones in the music scale. As a metal brush slides through a slot, an electrical contact is established which signals a particular power transistor to conduct a current through a relay coil. The relay, in turn, allows a storage condenser to discharge through a long electric line into the surrounding subterranean area, firing a plunger at high speed and sounding a stalactite.

The plungers that strike the stalactites have to be of magnetic material. Large ones are of soft iron with a brass tip sheathed in rubber. Small, rubber-tipped metal plungers are used for the smaller rocks and are mounted on Monel metal brackets. The relationship of plunger weight to stalactite size is crucial to tone quality. A large resonant rock will not respond properly to a small plunger, and a large, heavy plunger must be used to impart the proper vibratory motion to a large stalactite. These large plungers require heavy, electrically welded custom-made brackets. Rubber cushioning is used to eliminate the transfer of mechanical sound from plunger to mounting brackets. Brackets are installed in different ways, depending on the structure of an adjacent rock. Sometimes the Monel metal is wrapped around the adjacent rock, but more often the brackets are attached rigidly to the rock by lag screws. Resonant rock must be avoided for mounting brackets because they give off tone (an undesirable thump) as the plunger springs back into position.

The central point of the Stalacpipe organ is the terminal board and control panel, to which wires are directed from each octave block, the automatic player, stalactites, and the four-keyboard and pedal-board console. Any stalactite can be sounded automatically at a predetermined rate for testing. The volume can also be set at any desired level. Sprinkle says the console wiring can be expanded to five octaves and use every stalactite in the caverns that can be tuned. With precision sanding, the stalactites are tuned precisely to concert pitch with the key of A standardized at 440 cycles per second. Maintenance, thanks to factors such as conservative engineering, rugged construction, and steady temperature is at a minimum.

To secure a dry environment for the driving mechanism and electronic apparatus, four calrod units eight feet long have been installed in pipes in the top of the concrete and cinderblock platform on which each octave-block rests. The calrod units, similar in principle to those used in electric stoves, operate at half their rated voltage and without a thermostat. This low-level heat keeps the surface so dry that dust can be blown off the platform. For the console platform, eight calrod units are used.

Economy in the size of the octave-block power supplies is obtained by continuously charging condensers, the modern equivalent of Benjamin Franklin's Leyden jar. Each one is momentarily disconnected from the power supply just prior to propelling a plunger, and a moment after discharge, one of the larger plungers probably develops between one and two horsepower.

Miles of wire are used for stalactites now linked to the console. A special plastic-covered wire is used whose insulation will stand 1000 volts without breaking down. An all-pervasive stereophonic effect is achieved by the musical tones sweeping in through the Cathedral and surrounding area. No other instrument in the world has more area in which to produce its majestic music. An ethereal, undulating effect is achieved as harmonics reverberate through this underground system.

The console of the Stalacpipe Organ was especially constructed to specifications by the Klann Organ Supply Company of Waynesboro, Virginia, to meet the peculiar needs of this subterranean installation. It has five expression pedals, draw knobs with numbers rather than names to designate controls, and numerous couplers. Its special organ division names are "Pedal," "Harmonic" for softer effects, "Cathedral" for added brilliance, "Solo," and "Echo" for far-distant effects.

On July 29, 1956, the first composition, "A Mighty Fortess Is Our God," was performed. Long, intensive trials were made to test the durability of an instrument built with great care and conservation to last for years to come. Also recorded and played seasonally are the 18th Century "Dutch Hymn of Thanksgiving," "Silent Night," "Christ the Lord is Risen Today," and many others, including "Shenandoah." Figure 9-24 shows the Great Stalacpipe Organ.

Andrew Campbell and Benton Stebbins knew that "a picture is worth a thousand words." However, the candlelight used to illuminate the cavern combined with slow film and early cameras prevented this natural wonder from being photographed. Leading newspapers and magazines gave glowing reports of the Caverns, and Stebbins, being a professional photographer, realized a solution had to be found. American initiative prevailed when Stebbins engaged his wife to make India ink drawings of the spectacular views. These were photographed in the Stebbins studio and reproduced as postcards, becoming the first promotional illustrations of Luray Caverns. Figure 9-25 shows one of the first pictures made of Luray Caverns by this technique over 100 years ago.

Saracen's Tent, shown in Fig. 9-26, is recognized by authorities as one of the finest cave formations in the world. Water seeping through crevices in the limestone roof form stalactites from the ceiling and stalagmites from the floor. Lime and other minerals dissolved by water from above caused these formations to grow. Luray Caverns is an active cave, which means that the formations are still building at the very slow rate of one cubic inch every 120 years.

One of the other fascinating features of Luray Caverns is the Wishing Well, shown in Fig. 9-27. Located more than 164 feet underground, it is helping dreams come true for many people of all ages who suffer from crippling diseases. During the years 1954 through 1976, the Wishing Well has yielded $142,000 in small change, all tossed into the waters by thousands of visitors. Periodically the well is drained and all coins are washed, sorted, counted, and banked. The proceeds are then used to further the work of national health organizations.

An indication of the growing popularity of Luray Caverns is the fact that 26% more money was thrown in the pool in the last five years than during the preceding five years. There are 88 coins tossed into the Wishing Well for every 100 visitors to the Caverns.

In the past, funds have been given to the March of Dimes, the American Cancer Society, the American Heart Association, the National Society for Crippled Children and Adults (Easter Seals), Mus-

Fig. 9-24. The Great Stalacpipe Organ is the result of many years of research and work. (courtesy Luray Caverns)

Fig. 9-25. One of the first visual depictions of Luray Caverns. (courtesy Luray Caverns)

Fig. 9-26. Saracen's Tent is one of the finest formations in the world. (courtesy Luray Caverns)

Fig. 9-27. The Wishing Well has yielded great sums of money which is donated regularly to national health organizations. (courtesy Luray Caverns)

Fig. 9-28. Many stalactites and stalagmites are found in Luray Caverns. (courtesy Luray Caverns)

Fig. 9-29. These formations have fused together to form massive columns. (courtesy Luray Caverns)

Fig. 9-30. Another view of the magnificent formations in Luray Caverns. (courtesy Luray Caverns)

Fig. 9-31. Some of the formations present a white appearance. (courtesy Luray Caverns)

Fig. 9-32. The Igloo is one of the many flowstone formations seen in Moaning Caverns. (courtesy Moaning Caverns)

cular Dystrophy, and other non-profit groups. In 1976, the National Multiple Sclerosis Society received a check for $32,000 to assist in finding the cause and cure for this disease. The remaining money, about $22,000, was divided among other health organizations.

At its deepest point, Luray Caverns is more than 164 feet below the surface of the earth. The caverns were formed by water moving through pores and cracks in water-deposited rocks, a carving process that began many thousands of years ago. As water filled all the openings and dissolved the rock, the openings were enlarged. Eventually, rooms and passageways were shaped out. Some of the stalactites and stalagmites in the caverns are estimated to be 10 to 12 million years old. One of the most scenic spots in the mile tour is a deep lake that reflects the myriad colors and spectacular shapes of surrounding formations. Years ago, visitors began spontaneously throwing coins into the still waters of the pool. Although any amateur psychologist can predict that water attracts pennies the way honey attracts bees, no one can explain why cavern visitors chose these particular waters for coin tossing. Dream Lake, nearby, is a large, more impressive body of water.

To keep the water, normally six to eight feet deep, from overflowing the brick and concrete path which winds through the caverns, the well must be drained periodically and the coins removed. This is a tedious, hand-labor job. First, the water must be pumped from the pool and emptied into another pit. Then maintenance men move in with shovels, buckets, wheelbarrows, and brooms. The brooms are used to sweep down the craggy walls, to which coins sometimes cling. After the coins are separated and bagged, they are taken to a bank, where a special account has been set up for the money.

Figures 9-28 through 9-31 show some of the more spectacular formations in Luray Caverns. In addition to tours of the Caverns at Luray, there is much to see and do in the surrounding Shenandoah Valley. On the grounds is a historic car and carriage caravan, which is a collection of 75 antique vehicles, ranging from a Royal Coach from Portugal built and used prior to 1625 to autos of the 1930s. The Caverns Country Club Resort, a half-mile west of the Caverns, offers an 18 hole golf course overlooking the Shenandoah River as well as four hard-surface tennis courts. Also, one mile west of Luray Caverns is the Luray Caverns Airport, which provides free transportation for fly-in visitors between the airport, motels, and the country club. As can be seen, a complete vacation can be planned around Luray Caverns which should be pleasing for the entire family. Further information can be obtained by writing to Luray Caverns Corporation, Luray Virginia, 22835.

MOANING CAVERN

Located in Vallecito, California, the heart of California's historic Mother Lode, Moaning Cavern was first explored by gold miners in 1851. However, the cavern was long known to prehistoric people, who held it in awe because of the moaning sound emanating from the entrance.

On a professionally guided tour, the visitor descends through well-illuminated marble passages to the first level, where the true magnificence of this gigantic natural wonder is first revealed. The main chamber, large enough to hold the entire Statue of Liberty with room to spare, is quite beautiful. Strange shapes and figures adorn the ceiling and walls, making this a veritable treasure house of natural history. Visitors see the old entrance through which over 100 prehistoric people fell to their deaths and where early explorers, searching for gold, daringly lowered themselves into the vast cavern with long ropes.

From the first level, an all-steel spiral staircase descends 100 feet to the floor of the main chamber where visitors stand upon the spot where scientific excavations have revealed that more than 13,000 years ago, the bodies of prehistoric men fell or were thrown into the blackness of this huge cavern. The bones, preserved to the present time by the deposits of mineral-bearing waters, are thought to be the oldest human remains found in America.

As visitors travel through the cavern, a guide explains the 300 million years of geologic history and speaks of the explorations that have traced the cavern downward for 410 feet, and of the new explorations in progress. Moaning Cavern is awesome, fascinating, and very worthwhile to visit. Figure 9-32 shows the flowstone formation called the Igloo. Further information can be obtained by writing to Moaning Cavern, P.O. Box 78, Vallecito, California 95251. A tour takes approximately 45 minutes, and the cavern is open year-round except Christmas Day.

BOYDEN CAVERN

Boyden Cavern is located in the deepest canyon in the United States, the spectacular 8000-foot deep Kings River Canyon. The cavern is between Grants Grove and Cedar Grove of Kings Canyon National Park. The parking area and attractive building where information, tickets, snacks, and gifts may be purchased is ten miles before Cedar Grove where Highway 180 crosses the Kings River.

Boyden Cavern lies beneath the massive 2000-foot high marble walls of the famous Kings Gates. The entire trip through the cavern takes about 45 minutes, including a five-minute walk from the parking area. Deep within, the visitor will find the temperature a constant and refreshing 55° F. As visitors follow the well-lit and handrail-equipped trail, the guide points out and explains the many natural varieties of formations, massive stalagmites, columns, delicate stalactites, and splendid arrays of formations that defy description. Yet this beautiful and interesting cavern is only a small part of an extensive, five-mile long cavern system that formed as the marble was slowly dissolved and redeposited by underground water over a 300,000-year period. The water still tumbles from a small fissure, finds its way through the tortuous passage, and then disappears again. Overhead, thousands of tiny water drops shimmer on the still-growing formations.

National Park and Forest campgrounds are available near at Grants Cove, Indian Basin, Hume Lake, and Cedar Grove. Nearby sights are Grizzly Falls, Roaring River Falls, and the famous Zumwalt Meadows. The area also boasts the most spectacular hiking trails anywhere, and horses are available. Boyden Cavern is open year-round, and further information may be obtained by writing to Boyden Cavern, Box 817, Kings Canyon National Park, California 93633.

CAVE CITY

Cave City is highly unusual in that it is actually two caverns in one. Located in Vallecito, California, near Moaning Cavern, it offers the typical antiseptic guided tour for the average family, as well as a five-hour Wild Cavern Trip which is also guided but takes explorers through the muck and mire associated with exploring wild caves. Each year, millions of people tour commercial show caves, but some of the more adventurous look beyond the lights and handrails and wonder what it would be like to explore the unimproved regions.

Located in the heart of the Mother Lode, Cave

City was once a roaring and rich gold mining town where miners and visitors from all over the world took time from their frenzied search for gold to visit and explore what was then called Mammoth Cave. It was the most well-known landmark in Northern California and was visited by nearly everyone passing through the area. The cavern was also used for a time as a hideout for the Joaquin Murrieta gang, and there are stories of loot hidden in the cavern. As gold mining declined, Cave City fell to ruin and the cavern was sealed and forgotten. Today, little sign exists of Cave City—just a dam with water thundering over, broad meadows with a meandering stream, and the slowly advancing forest.

Recently, new discoveries within this extensive system have expanded it to three times its original size. New discoveries are being made almost daily, including huge rooms six and eight stories high, lakes over 200 feet deep, and spectacular crystal formations. Cave scientists estimate that less than a third of the cavern is explored.

So extensive is this cavern system that two trips are offered. The first is a standard family tour through historic Mammoth Cavern, which offers improved trails, guardrails, and electric lighting. The second trip (the one most interesting to spelunkers) is called the Wild Cavern Expedition and is offered only to healthy men, women and children aged twelve and over. This latter expedition is composed of a maximum of ten people, who are guided by two experienced leaders. A typical trip expends 4½ to 5 hours and involves crawling, climbing slopes with ropes and ladders, and boarding rafts to cross deep, crystal-clear lakes. New passages and rooms that have only recently been discovered are offered for exploration, and there are a large number of crystalline formations which are truly magical in appearance. Group leaders relate rich historical and geological information while giving aid and instruction to the group members as they make their way along the challenging route.

For the Wild Cavern Expedition, hard hats, lights, coveralls, gloves, and other special equipment are provided for the $45 fee. Explorers must bring loose-fitting heavy shirts and pants, towels and personal items. A change of clothes is also recommended, along with food for the five-hour trip. All members of the trip are required to sign a waiver of liability.

The five-hour tour technically takes place in Quill Lakes Cavern, which is a part of Cave City. Delicate crystalline formations, rare minerals, and clear green lakes, some of them over 200 feet deep, are all encountered. Four groups of ten make up the normal daily routine. Spelunkers are welcome, but no unguided tours are offered. All exploring equipment (ropes, ladders, rubber rafts, etc.) are provided for the price of the tour.

For the wild trip, it is advisable to make reservations well in advance by calling Cave City at (209) 736-2708. A 25% deposit is required for such reservations. Spaces are limited and summer weekends are often booked three to four weeks in advance. Cameras are not generally permitted on this trip, as Cave City notes that they do not normally survive the trip. However, it should be possible to talk the leaders into allowing you to bring a camera along as long as you can demonstrate that you have had previous experience with photographic equipment in wild caves. As previously mentioned, spelunkers are welcome, and special discounts may be in effect for clubs that book reservations in advance. Further information may be obtained by writing Cave City Expeditions, P.O. Box 78, Vallecito, California 95251.

Chapter 10

The National Speleological Society

The National Speleological Society is a non-profit organization affiliated with the American Association for the Advancement of Science. It was founded for the purpose of advancing the study, conservation, exploration, and knowledge of caves. In addition, the Society collects and publishes information relating to speleology in this country and all parts of the world. Membership in the Society encourages protection and conservation of caves and provides contact with thousands of serious cave explorers and scientists in the United States and foreign countries.

ORGANIZATION

The National Speleological Society is administered by a Board of Governors consisting of twelve Directors and four Officers who meet about three times annually to consider matters of policy. Four Directors are elected annually for three-year terms. Each year the Directors elect the President, Executive Vice President, Administrative Vice President, and Secretary-Treasurer. These officers are responsible for the management of the Society under the policies adopted by the Board.

There are more than 100 local chapters of the National Speleological Society throughout the country. These local groups, usually known as *grottos*, conduct regular meetings and serve as a means of organizing speleogists within a geographical area. Members are urged, but not required, to join a local grotto in order to take part in joint explorations and publication work and to enjoy the programs and fellowship of other members.

Upon request, the Society will provide information concerning the formation and operation of new grottos or other internal organizations, i.e., Regions, Sections, and Surveys, and Conservancies.

RESEARCH SUPPORT

The Ralph W. Stone Research Fund, named in honor of the second President of the Society, provides grants in support of individual research in the field of speleology. Among these is an annual $1000 award to a graduate student whose thesis deals with a speleological topic.

The formation of Society Projects is encouraged, and money grants are made to assist in obtaining equipment or publishing findings. The designation of a National Speleological Society Study Group has been created for very long-range projects.

PUBLICATIONS

The *NSS News* is published monthly and contains accounts and photographs of current explorations, business matters, grotto news, book reviews, and abstracts of current articles of interest. The *NSS Bulletin* is devoted to reports of original investigations on speleological topics. The *Bulletin* is responsible for the respect the Society now holds in the scientific world. These publications are received by all members of the Society.

LIBRARY

The NSS Library is located in Huntsville, Alabama (Cave Avenue, 35810). It is supervised by a professional librarian, contains over 800 volumes, and receives 140 publications from many countries. Complete sets of all grotto publications and other periodicals pertaining to caves are available here. This is the most complete reference library dealing with caves in the United States. Members may write for a list of circulating titles.

ACTIVITIES

The NSS Annual Convention is held for a period of one week in a section of the country noted for its cave features. Field trips are held the first three days, with scientific papers, symposia, demonstrations of techniques, exhibits, a Congress of Grottos, and a banquet occupying the rest of the week. Many members camp, but other housing is also available. Recent conventions have been held in Birmingham and Huntsville, Alabama; Springfield, Missouri; Lovell, Wyoming; State College, Pennsylvania; Blacksburg, Virginia; Elkins, West Virginia: and Bloomington, Indiana.

The members and grottos of the Society frequently sponsor weekends of cave exploration, training sessions, and opportunities to get together for the enjoyment of discussing past caving trips and the planning of future activities. The national Society designates NSS Field Trips, Expeditions and Speleological Seminars that are open to all members. Week-long "Speleocamps" have recently been introduced for the enjoyment of contributing to intensive studies of cave systems.

CONSERVATION

Caves are unique. The increasing exploration of caves necessitates measures to protect the geological and biological rarities found underground. All members are expected to pledge that they will do nothing that will deface, mar, or otherwise spoil the natural beauty and life forms in caves. Members can aid in this conservation program by explaining to all those concerned with caves the absolute necessity of preserving caves intact.

The Society has recently been active in wilderness reviews of federally-owned cave areas. Through the contribution of many members, Shelta Cave in Huntsville, Alabama, and the Trout Rock Caves near Franklin, WV were purchased as natural preserves and important biological study areas. Members can become directly involved in conservation efforts through the many NSS Task Forces that are directing their efforts toward protecting particular areas or solving particular conservation problems.

Individuals contributing more than $100, or more than $1500, to support the Society's efforts in conservation and other areas are designated as Conservators or Benefactors of the Society, respectively.

The National Speleological Society believes: that caves have unique scientific, recreational, and scenic values; that these values are endangered by both carelessness and intentional vandalism; that these values, once gone, cannot be recovered; and that the responsibility for protecting caves must be assumed by those who study and enjoy them.

Accordingly, the intention of the Society is to work for the preservation of caves with a realistic policy supported by effective programs for: the encouragement of self-discipline among cavers; education and research concerning the causes and prevention of cave damage; and special projects, including cooperation with other groups similarly dedicated to the conservation of natural areas.

All contents of a cave (formations, life, and loose deposits) are significant for its enjoyment and interpretation. Therefore, caving parties should leave a cave as they find it. They should provide means for the removal of waste; limit marking to a few small and removable signs as needed for surveys; and especially exercise extreme care not to accidentally break or soil formations, disturb life forms, or unnecessarily increase the number of disfiguring paths through an area.

Scientific collection is professional, selective and minimal. The collecting of mineral or biological material for display purposes, including previously broken or dead specimens, is never justified, as it encourages others to collect and destroy the interest of the cave.

The Society encourages projects such as: establishing cave preserves; placing entrance gates where appropriate; opposing the sale of speleothems; supporting effective protective measures; cleaning and restoring over-used caves; cooperating with private cave owners by providing knowledge about their cave and assisting them in protecting their cave and property from damage during cave visits; and encouraging commercial cave owners to make use of their opportunity to aid the public in understanding caves and the importance of their conservation.

Where there is reason to believe that publication of cave locations will lead to vandalism before adequate protection can be established, the Society opposes such publication.

It is the duty of every Society member to take personal responsibility for spreading a consciousness of the cave conservation problems to each potential user of caves. Without this, the beauty and value of our caves will not long remain with us.

SERVICES

All members have use of the NSS Library and access to the Cave Files. They may borrow numerous slide programs from the audio-visual aids library and also certain items of speleological equipment. They receive lists of publications that may be purchased through the Society bookstore, often at a member discount. They may participate in all of the many Society activities and have access to information from special Society committees for Safety, Conservation, Research, etc. Probably the most important "service" to members, however, is the association with fellow cave enthusiasts, provided by Society membership.

CAVING COURTESY

Chapter 6 dealt entirely with caving courtesy. However, you really can't say too much about this subject. NSS prints a very handy pamphlet on caving courtesy, and a portion of it is presented here, as it includes a few more details that were not directly covered in Chapter 6.

The people of the surrounding community as well as the individuals who control the cave entrance and its access routes may feel affected by cavers. Almost anybody can get a cave closed if you stimulate him to do so, so keep in mind the fact that fouling up laundromats with mud and raunchy good fun in the restaurants can do just as much harm as horse-tail-pulling and corn theft.

People rarely own or lease land they do not care about. Regardless of the rumors you hear or the current practice in a certain place, regardless of the owner's ability to observe you entering the cave, *ask first*. Disrespect really is an irritant. Before entering any cave, be sure you have the permission of the owner. Even when a "standing invitation" has been extended, you should contact him before each visit.

Many caves are situated on public or commercially owned land. There will be a management of one sort or another and possibly a policy toward caving. Caves are controlled and frequently closed on federal and state lands. You should contact a ranger or superintendent and adhere strictly to their regulations. Cavers should expect to be asked to in some way show qualifications to enter difficult caves. Further, you may have to prove some interest other than recreation in some cases. The eventual opening of closed public property caves depends partially on our cooperation now. On Indian reservations, frequently you will not be allowed to cave. Seek out an assistant to the Chairman of the Tribal Council for permission. On commercially owned property such as quarries, logging areas or the like, you can expect that management will dissuade caving. Cooperation and friendly dialogue may eventually lead to a change in policy. Sneaking in will lead to an adamant closure.

Regional situations vary, but in general, one of the biggest problems is that so many owners are rural people and many out-of-town cavers are from urban centers. "City" people frequently know little of what to expect from livestock in the way of behavior, nor do they instinctively know how to avoid damage to crops and fields. Care and consideration on the part of the spelunkers will result in warm invitations to return. However, when an owner feels that caving is a threat to his property, livelihood, or community standing, he will close his caves. One of the best ways to foster good relations and attune yourself to rural caving is to stop to talk awhile with the owner and his family.

While visiting with the owner, keep in mind some of the things you want him to know about yourself, as well as those you must learn about him. You should put the owner into your safety picture with information on your whereabouts and who he can contact should problems arise. In this vein, expect that due to publicized accidents, owners in heavily cave areas may be apprehensive about your visit. In some areas, it will be to your advantage to carry release or liability waver forms. You should make a point of learning any laws in his area which protect him from liability during your visit. You should try to find out exactly to whom you are speaking. In this way you gain a name for your Christmas card list and know how to pronounce it. Further, you will gain an insight into the family structure hinted at by the names on the mailboxes. Possibly you will hear of a neighbor's new sinkhole. Possibly you will hear of problems your host is having with cavers. Possibly some fantastic cave lore will be related to you. Needless to say, attention paid to the problems will pay off both in the short and in the long haul. Before leaving the owner, inquire about such things as what to do with your car and how to reach the cave, and possibly, camping.

You need a car to get there, but farmers are constantly moving machinery or livestock about, so check on parking to be sure your car will not block a lane that is in use. Drive on the existing trails and roads to prevent rutting, scarring, and erosion in fields. This is particularly important on hills and in wet weather, as a slight spin of the wheels can lead to problems maintaining the road after it ruts. By all means, remember your host does not really enjoy towing cavers' cars out of muddy dirt roads or fields, and he probably prefers that you change your oil in town.

Eventually, you will get out and walk. Keep any children with you out of the sheds and off the machinery. When you cannot avoid climbing a fence, do so at its strongest post and make sure that you do not leave sagging wires or missing rails. Ranching and farming people don't consider chasing animals fun. If there are grain fields, the owner will appreciate it if you ask which route he'd prefer you use to reach the cavern. Walking through grain fields can result in permanent damage to the crop.

Gates are frequently encountered in rural areas. The owner expects you to close and securely fasten any gates you pass through. Again, in speaking with him, you may learn of some he prefers open on a given day. Entering and leaving a cave, replace any barriers that may have been put there to keep animals out. In warm weather, the cool air and water in the typical entrance will attract cattle. Around the entrance, you may see piping and pumps. If cave water is being used on the farm for irrigation or drinking, you can expect that its quality is a sore point with the owner. You must use a great deal of care to avoid muddying or spoiling his water supply.

An owner frequently is familiar with his cave and is interested in conserving it. His interest in its lore, wildlife, and formations may date from his own youthful explorations. While telling him of new discoveries will interest him, showing him how to protect his bat population or shouldering bags of extracted trash will please him more. While enjoying a cave, remember that the owner or his relatives or friends may visit it next. It is in your own best interest that you leave nothing inside a cavern that does not naturally occur there. Litter, carbide, food, plastic wrappers, photographic debris, names or hometowns sooted on walls or formations, all detract from the natural appearance of a cave. Further, many unlikely items will wreck the life cycle if allowed to contribute food or poisons into the delicate balance found there. Cave fauna are extremely hard-taxed just to survive and should be allowed to remain in their natural habitat undisturbed. Probably all cavers realize they should not "collect" or vandalize formations. However, few of us can manage to keep the accidental breakage under complete control. Remember, a single nudge of your hard hat or helmet will destroy practically any small formation. Where you have an owner who is interested in his cave, you have an individual who will be upset at its deterioration.

After exiting, cavers have on rare occasion dumped spent carbide around cattle. Not only is this a form of littering, but most ranchers and farmers regard carbide as poisonous, and livestock deaths have been blamed on cavers. All carbide and spent carbide must be removed from the premises.

Work involving surface mapping, measuring, or electronic equipment could upset an owner who had not been briefed on its uses. Cavers using fluorescein dye for water course tracing should first talk with area residents. Once the tap water has turned green, a tardy "It's harmless" will not improve your caving prospects. Perhaps avoiding an officious or hurried demeanor is one of the best ways to show an owner that your visit is for innocent recreation. If you do get a report or map published, by all means see that your cave owner gets a copy.

After caving, we all feel grungy and groady. But don't mess up your landowner relations by leaving local public restrooms muddy. In the woods, do not use soap directly in a body of water. Do your soaping and dishes in a bucket and discard it at least fifty feet from open water.

Keep in mind the nearest public restrooms and plan accordingly. In the field, use whatever sanitary facilities the owner has already provided. If the need is great and facilities nil, choose a spot at least fifty feet from open water and bury extrement in the topsoil layer no more than six to eight inches deep.

Far from home you will need a place to stay overnight. Why run the risk of offensive camping manners messing up your "in" at the cave? If you can reasonably go elsewhere, never camp at the cave. But if you must camp and are granted permission to do so, stick to the golden rule.

Use whatever campsite exists already. Don't build structures, drive nails, break branches, or otherwise deface the camp. Any trash, even small items such as cigarette butts, are potential irritants to a sensitive owner. Use a small stove for cooking, since it leaves no mark. Toasting marshmallows means specific permission for a fire. Keep a campfire small, since a brush fire could make cavers unpopular locally. Use dead wood instead of maiming live trees. When you are finished, check for live sparks by mixing copious amounts of water into the ashes with a bare hand. Leave the camp cleaner than you found it. Idealistically, a campside is so clean as to disappear into its surroundings. Try to leave the landowner happy with your visit. After all, he need not always give cavers access to the caves we enjoy.

CAVE DIVING

Underwater exploration of inundated caverns was alluded to in Chapter 4. The National Speleological Society has its own Cave Diving Section, which offers a great deal of information and even training to those spelunkers who might wish to further their pursuit by donning scuba gear. This is a highly specialized field and one which is potentially lethal (to a high degree) when proper training as to technique, equipment, and safety has not been gained ahead of time. NSS has gone a long way to train those interested in cave diving and has certainly contributed to the utmost to further this branch of spelunking.

Many of the most interesting features of a cave can be found within the cavern, that area of the cave which receives surface light. The objective of the

NSS diving course is to introduce the student to the cave environment under the supervision of an experienced cave diving instructor. Lasting a single weekend, the course includes lectures on diving philosophy, the cavern environment, practice of safety procedures, and three cavern diving sessions. Upon satisfactorily completing the course, the student is awarded a cavern diver card by the NSS Cave Diving Section.

The Cavern Diver course is the first step in the four weekend-long cave diving instruction programs offered by the Section. From the time of completion of the cavern course, the student has two years to complete the full cave diving course, if he so desires. This time allotment is to allow divers living some distance from the springs to more easily obtain training.

For those cavern divers who wish to more fully develop their diving skills, the Section offers a full cave diving course. It is divided into two modules: Basic Cave Diving and Cave Diving: The basic course lasts one weekend and emphasizes those skills necessary to dive that portion of a cave accessible on a single-tank air supply, including line laying, swim technique, buddymanship, emergency procedures, and decompression. The Cave Diving course lasts two weekends and provides the student with a variety of cave diving experience under many different conditions, upon which the student can build safe diving habits. Additionally, other organizations, such as PADI, NAUI, YMCA, and NACD offer instruction in cave and/or cavern diving. These organizations can be contacted for further information on their courses.

The following equipment is the minimum necessary for safe cave diving. Unless you have each of the following items in good working order and are thoroughly familiar with their use, you cannot begin to cavern/cave dive safely.

- ☐ Mask, fins, wet suit.
- ☐ Single 72 cubic foot cylinder or larger, filled to at least 1500 psig.
- ☐ Single hose regulator equipped with an additional second stage and a submersible pressure gauge.
- ☐ Two dependable underwater lights.
- ☐ Front or back mounted buoyancy compensating device with an automatic inflator.
- ☐ Dive knife, slate, and pencil.
- ☐ Watch, depth gauge, and diving tables.
- ☐ Nylon guideline, 3/32 to ⅛ inch, on reel.

For cave diving beyond basic cave diving, each diver needs the following additional equipment:

- ☐ Dual tanks with dual valve manifold (crossover bar or dual single tanks are not acceptable).
- ☐ Dual single hose regulators (one for each tank valve), one with a submersible pressure gauge, the other with automatic inflator.
- ☐ Watch, depth gauge(s), dive tables and slate and pencil.
- ☐ Forearm knife and line clips.
- ☐ A third underwater light (one of the three must be as bright as a 30-watt light or brighter and have a burn time of at least 50 minutes).

An analysis of Florida cave diving accidents has shown that at least one of the following three safety procedures was ignored in every incident:

1. Always reserve ⅔ of your starting air supply for the trip out of the cave. This additional air will permit time for you to handle emergencies. This rule applies best to teams utilizing similar tank configurations and air pressure. Do not begin a dive with less than 1500 psig and monitor your submersible pressure gauge constantly.

2. Always run a single, continuous guideline from the cave entrance throughout the dive. Secure the line first in open water and again within the cave entrance to prevent its removal by swimmers or open water divers.

3. Avoid deep diving in caves. The average depth for fatal dives in which the first two procedures were followed was greater than 150 feet. Confine your cave dives to depths less than 130 feet, and cavern dives to 60 feet.

Most cave diving accidents are the cumulative result of several safety violations in addition to those listed above. There, the following safety procedures should be rigorously observed on all cave dives:

1. Become proficient in emergency procedures by practicing them in open water prior to the

dive. Practice buddy breathing with each new team member, and before the first dive of the day with each buddy. Practice other emergency procedures regularly. This offers both physical and psychological advantages.

2. Avoid panic by knowing and observing your own limitations and by building up experience slowly. Never depend on another diver's ability to carry you through the first dive. Practice self-rescue regularly. The dive must not exceed the limitations of the least experienced diver.

3. Avoid silt by using your buoyancy compensating device to stay near the cave's ceiling. Check behind yourself frequently to insure silt is not being stirred up.

4. Stay within arm's reach of the guideline and maintain visual contact with it at all times. If visibility is poor, maintain hand contact, but do not pull on it.

5. Avoid passageways you can not turn around in easily.

6. Stay near your buddy at all times, and watch for his signals. Should difficulties develop, notify your buddy immediately and exit the cave together.

7. Remember, anyone can cancel the dive at any time, for any reason. All divers must head back out together.

Your safety depends in large on the good judgment you use in planning the dive and following the above safety procedures. If you observe your limitations, plan the dive within these boundaries, and follow these safety procedures, you can enjoy a safe and productive dive. To ignore these procedures is to risk your life unnecessarily and your buddy's as well.

THE CAVE ENVIRONMENT

No amount of previous open water experience can adequately prepare you for cave diving. Regardless of their previous open water experience, most cave diving victims were untrained in cave diving procedures, inadequately equipped for the planned dive, and/or making one of their initial cave dives. Many were extremely inexperienced in other types of diving, and no less than 15 were certified scuba instructors who had no cave diving training.

Why did these divers drown? The answer lies in part with their ignorance of the unique hazards found in caves and their failure to recognize and deal with these hazards adequately.

For example, when cave diving, the ceiling restricts direct access to the surface, making you more dependent upon your equipment and its proper function. Should an emergency such as an air failure occur, you must exit the way you came in—out and then up. Yet many divers, unaware of this possibility, fail to plan for such an emergency.

Many divers rely on their dive light and memory of the cave to navigate the cave's maze-like passageways. Should their dive light fail (probably the most common cause of cave drownings) or their swimming technique stir up silt, reducing visibility to near zero, there are only two things which will help them exit safely: having practiced emergency procedures (reducing the panic factor), and the safety guideline connected with the surface.

Yet despite these hazards, thousands of cave dives are made each year in complete safety by those who have learned to cave dive properly. In its first year, more than 110 Abe Davis Safety Awards, named in honor of an American cave diving pioneer, have been awarded by the NSS Cave Diving Section to divers completing and logging 100 safe cave dives.

For these trained cave divers, the day's conventionality is put behind as they move through gin-clear, 72° water, enjoying scenic vistas and the quiet beauty of huge cathedral-like rooms safely. They are divers much like anyone else, differing only in that they have learned about the quiet, strange and beautiful environment of the underwater cave. As a result, they have acquired the necessary training and safety equipment, and on every dive, rigorously follow a set of proven safety procedures developed over many years by experienced cave divers.

The best way to become a safe cave diver is to first become a certified scuba diver and accumulate a sufficient amount of open water diving experience to fully develop your diving skills. An excellent way to do this is to take an advanced diving course. Do not go into cave diving without first acquiring cave diving training. The majority of cave diving victims have made fewer than five cave dives. Cave diving can be considered purely an equipment sport only under ideal conditions. Be trained for the not-so-ideal conditions.

STRUCTURE OF THE SOCIETY

As is obvious from the previous discussion in this chapter, the National Speleological Society is not some hole-in-the-wall organization made up a few good ol' boys who like to explore caves now and then. The NSS is a highly structured organization which has always answered the static and ever-changing needs of the amateur spelunker. Figure 10-1 shows the overall structure of the society. As you can see, it includes many branches, sub-branches, and divisions.

To a greater extent than in most other organizations, the NSS is based upon individual member participation. It is not uncommon for the Society to have difficulty finding a volunteer to fill a position. Inevitably, almost anyone who is interested in a given area and willing to work can find a committee to work on within the Society. Because of turnover in the leadership of many committees, the qualified and interested committee member may soon become the chairman.

There are many cave-related activities for which there is no organization or committee currently responsible. The individual member interested in pursuing one of these activities need only volunteer to do it. There are many opportunities to conduct a cave survey, develop and execute a management plan for a specific cave on behalf of its owner, resolve a landowners relations problem, perform a conservation project, or raise funds to purchase a specific cave. All it takes is individual interest—the organizational mechanisms of the NSS, and its normal laissez-faire approach, allow for accomplishment of the project within the Society.

Members of the NSS are encouraged to join one or more internal organizations. There are four types of internal organizations: chapters (usually called grottos), regions, sections, and surveys.

GROTTOS

About 200 local chapters, or grottos, located around the country represent the bulk of the internal organizations of the NSS. They conduct regular meetings and primarily serve the purpose of cavers within their general area. Members of grottos survey caves, install and maintain cave registers, build and maintain cave gates where this becomes necessary, and hold cave clean-up campaigns. Many grottos publish as a regular part of their program. Grotto publications vary from simple newsletters to elaborate, professionally done periodicals which alone are worth grotto dues. Grottos may publish cave maps, and newsletter contents are often highly informative, sometimes technical, and always of interest to area cavers. It is not unusual to find regional news published in grotto newsletters.

Most grottos, to one degree or another, maintain caver training programs and keep up to date on the latest developments in safety, equipment and technique. Grottos are often involved in cave conservation programs. Some grottos maintain cave rescue teams, and many maintain libraries and equipment stores.

While NSS members are not required to join grottos, they are urged to do so, for the grotto is truly the backbone of the NSS. Even though a group of cavers comprising a grotto may do nothing more than "go caving" for sport, the grotto still provides fellowship among cavers and provides responsible persons with whom to cave. Information on forming a new grotto where there currently is none is available from the chairman of the NSS Internal Organizations Committee.

REGIONS

Frequently, a group of grottos within a general geographic area will unite in a form of loose association called a region or regional association. Individuals within this geographical area who are not associated with any grotto may also belong to the region.

It is rather difficult to define a region, since the

structure of each of the regions within the NSS is different. Some regions are formed principally for the purpose of fellowship among the grottos; others have specific organizational purposes, such as training programs and coordination of caving activities. All of the regions give grotto members the opportunity to formally get together to trade information and solve problems common to the entire geographical area. Region meets or conventions, then, seem to be a principal purpose of regional organization. Some regions publish and may, in fact, serve as the principle publishing organ for a group of grottos. Region seminars on safety, first aid, rescue procedures, and conservation matters are frequently held through the collective efforts of the several grottos within a region, for the benefit of all members within the region, where it might be otherwise impossible for an individual grotto to muster the forces and talents for such programs.

Whatever may be the various reasons for regional organization, the region provides the opportunity for cavers among several grottos to deal with their own specific area needs. Currently established regions in the Society include:

☐ Arizona—Comprised of grottos in Arizona. Sometimes meets jointly with Southwestern Region.

☐ Mid-Appalachian—Holds a business meeting in midwinter with fall and spring caving weekends.

☐ Mississippi Valley-Ozark—Has gatherings in spring and fall, usually in Missouri. MVOR meetings feature speakers and promote fellowship and caving.

☐ North Country—Located around the Great Lakes. Publishes *NCR News*. Meets four times a year.

☐ Northeastern—Holds spring and fall meetings. Sponsors a rescue organization. Publishes *Northeastern Caver*.

☐ Council of Appalachian Volunteers Engaged in Speleology (CAVES)—Coordinates caving activities, particularly in conservation, in the central Appalachian area. Publishes a newsletter.

☐ Northwest—Includes some Canadian groups as well as U.S. grottos. Publishes *Northwest Caving*. Puts on a summer caving meeting and winter symposium. Active in cave management and conservation work.

☐ Ohio Valley—The newest region, publishes *Ohio Valley Caver* three times a year. Sponsors two rescue teams in Kentucky and maintains a call-down list.

☐ Rocky Mountain—Has annual meetings. Their publication, *Depths of the Rockies*, appears irregularly.

☐ Southeastern—Holds a winter business session and a summer Cave Carnival.

☐ Southwestern—Publishes *Southwestern Cavers* six times a year. Conducts two field meetings a year (Easter and Labor Day, usually), and a midwinter technical meeting with representatives of government, etc.

☐ Texas—*The Texas Caver*, which appears six times a year, is one of the better-known caving publications in the U.S. Texas cavers have a mid-September meeting and are active in several projects in Texas.

☐ Virginia—The oldest NSS region, started in 1950. Publishes a monthly newsheet, *VAR-FYI*. Recently published a book on its history (1950-1980). Operates the Cave Rescue Communications Network with a call-down list. Holds fall and spring meetings with business sessions and other activities. Has sponsored conservation and cave protection activities and maintains contact with other environmental groups.

☐ Western (California and contiguous states)—Their publication is the *California Caver*. Annual meeting held during Labor Day weekend. Sponsors regional seminars including topics such as caving techniques, safety, etc. Active in cave conservation.

SECTIONS

The National Speleogical Society is further broken down into nine different sections. This type of structuring provides a framework to bring together cavers with similar specialized interests. While a few sections may require specialized skills for active membership (cave diving, for example), subscription to a section publication are available to those without that skill but with an interest.

American Spelean History Association

The purpose of the American Spelean History Association is to study, disseminate and interpret spelean history—the folklore legends and historical facts on caves, cavers, and caving. Papers resulting from research are read at the annual NSS Convention and printed in the section's quarterly *Journal of Spelean History*.

Biology Section

The Biology Section of the NSS was organized in 1965 for the purposes of encouragement and facilitization or research and communications among biologists interested in the study of life in caves of the United States, Mexico, Central America, and Canada. The section publishes the *North American Biospeleology Newsletter* five times a year, sponsors a session at the annual NSS Convention for the presentation of papers, and holds one business meeting a year at the Convention.

Cave Diving Section

The NSS Cave Diving Section is the largest cave diving organization in the United States, with members in almost every state. While section members are very active in diving springs in Florida, they also dive mines and swamps in the northern states, conduct high-altitude sump diving in the West, perform motorized and stage diving in the South, dive sea caves in the Northeast, survey Bahama Blue Holes, and conduct studies of various natural springs in Mexico. The section is also active in the development of underwater rescue equipment, and sponsors a comprehensive cave diver and instructor training program. It also holds national technology transfer seminars twice a year and publishes *Underwater Speleology*.

Cave Conservation and Management Section

The purpose of the Cave Conservation and Management Section is to encourage cave conservation and provide a central clearing house for information, expertise, and research in the field of cave management. The section co-sponsors regular cave management symposia at national and regional levels and assists the NSS Cave Ownership and Management Committee in managing Society caves. Some of the symposium proceedings have been published and are available from the NSS Bookstore. Additional projects are the development of literature for the specific use of cave managers and people directly involved with caves, and the development of guidelines for the NSS Board of Governors to assist in decisions regarding cave acquisition.

Computer Applications Section

Those Society members interested in computers and their application to cave related projects recently formed the Computer Applications Section. The section conducts a workshop at the annual NSS Convention and publishes the *CAS Newsletter*. One major interest of section members is computer programs to process cave survey data and provide graphical representations of the cave in two and three dimensions. Recent issues of their newsletter contain papers on loop closure analysis.

Cave Geology and Geography Section

The NSS Section of Cave Geology and Geography includes cave and karst geologists and geographers, both amateur and professional. It sponsors a session at the annual NSS Convention for the presentation of papers publishes *GEO*^{2} triannually. This section also maintains the Society's Cave Map Symbols list and lists of the world's longest and deepest caves.

Social Science Section

The Social Science Section promotes the scientific study of the human aspects of caving through the disciplines of the social sciences. Most of the activities of the section during the past few years have consisted of presentation of papers at the annual NSS Conventions dealing with caving habits of American cavers, minorities in caving, and solo cavings. The section publishes the *Social Science Newsletter*.

Vertical Section

The Vertical Section publishes *Nylon High-*

way, and organizes the vertical session and prusik contest at each NSS Convention. The section sponsors a vertical workshop at each Convention which provides a forum for exchange of ideas on vertical caving techniques and equipment.

Women's Section

The Women's Section of the NSS was chartered in 1976. The women who formed the section wanted to provide a forum where cavers could discuss caving problems they encountered as women, where new techniques and equipment designed with a woman's physiology in mind could be demonstrated, and where support could be provided which would encourage caving women to maximize their abilities as cavers and as leaders in their grottos and in the NSS. This section publishes *Women Cave.*

SURVEYS

In almost every state containing a number of caves, there is a group of NSS cavers actively maintaining data on the caves and pursuing further exploration. Only a few of these have asked for NSS charters as Internal Organizations, which others—as well as some of these—are recognized NSS Study Groups and some may be independent. The best way to make contact with a survey if you want to become active in this type of product is by inquiry through your local grotto or region.

Some cave-rich states assign working groups to specific counties to hunt for caves on a systematic basis, for exploration, and to develop data including maps. Such an approach has been led to a dramatic increase in the number of caves known in the state. Discoveries of virgin caves are still being reported from well-known and frequently visited wild caves. In this particular area of work, armchair cavers gifted in developing good relations with cave owners may be as useful as hard-core explorers and specialists in processing cave data.

CONGRESS OF GROTTOS

Once a year, delegates to the Congress of Grottos from internal organizations meet at the annual Convention to discuss and vote on issues that concern the members or that the members feel should be brought to the attention of the BOG. The number of delegates for each IO is based upon the number of NSS members in that group. Unafiliated NSS members can form a group and also be represented at the Congress. Issues are solicited from internal organizations or any NSS member, and thus, the range of subjects is limited only by the interest of the members. The COG is an advisory body to the BOG. BOG is influenced by COG recommendations which are decisively supported and consistent with NSS goals. Issues may be submitted at any time to the chairman of the COG or the Issues Chairman of the COG.

BOARD OF GOVERNORS

Since its founding in 1941, the NSS has been governed by a Board of Governors. Since reorganization in 1960, the Board of Governors has consisted of the four officers and 12 directors. The President of the Society presides over the BOG meetings and does not vote except to break or make a tie. The BOG has full power to conduct and supervise all business of the Society.

The Board of Governors meets three times per year. One of these meetings is held during the NSS Convention. The other two meetings are held in various parts of the country on a rotation basis so that all NSS members will, at one time or another, find it possible to attend a BOG meeting. Rotating the location of the BOG meeting also gives board members a chance to see local activities and hear local issues.

Directorate

The twelve directors of the Society are called the Directorate. The directors are elected by the NSS membership. Each director's term is three years. Terms are staggered so that four directors come up for election each year. The directors are responsible for establishing the overall goals and policies of the Society. Another important job of theirs is to elect the four officers.

DEPARTMENT OF THE PRESIDENT

The Department of the President is divided up into a number of committees, each of which is responsible for a given area. Each will be discussed briefly.

AAAS Representatives

The NSS is associated with the American Associated for the Advancement of Science. The Society has two representatives to the AAAS, one in biology and one in geology and geography. They represent the NSS at AAAS section meetings and coordinate cave-related papers for presentation at the annual AAAS convention.

Awards Committee

Each year at convention, the NSS recognizes cavers who have made outstanding contributions to speleology or to the NSS. Awards are given by the Board of Governors on the recommendation of the Awards Committee and are presented during the convention banquet program. Anyone knowing of someone who should be nominated for an award should contact the Awards Committee chairman. Below is a listing of the various awards which are presented by this committee.

1. Honorary Member Award—One Honorary Membership may be given each year for outstanding contributions to the field of speleology. The Honorary Member Award is of equal stature to the Outstanding Service Award and together they constitute the Society's highest awards. The Honorary Member Award includes a life membership in the Society.

2. Outstanding Service Award—One Outstanding Service Award may be given each year for outstanding service to speleology and the Society. The Outstanding Service Award is of equal stature to the Honorary Member Award, and together they constitute the Society's highest awards. The Outstanding Service Award includes a life membership in the Society.

3. Certificate of Merit—Up to three Certificates of Merit may be awarded each year at the Convention for specific accomplishments in cave exploration, study, or conservation. Certificates may be given to individuals, jointly to no more than three individuals, or to organizations. Emphasis is on recent accomplishment. Certificate of Merit recipients who are, or become, members of Society will become Fellows of the Society.

4. Fellow of the Society—The title Fellow of the Society is an award given for service in the field of speleology, whether scientific, exploratory, or administrative. Emphasis usually is on continued service over a period of time. The total number of Fellows is limited to 20 percent of the Society membership.

5. Ralph W. Stone Research Award—The Ralph W. Stone Research Award is given to a graduate student preparing a thesis on a speleological subject. The recipient is chosen by a subcommittee of the Research Advisory Committee based on nominations or applications received by the Committee. The award consists of $1,000 from the Ralph W. Stone Research Fund. The recipient is required to give credit to the NSS when the work is published and donate a copy of the publication to the NSS Library.

6. Lew Bicking Award—The Lew Bicking Award is given to recognize an individual Society member who, through specific actions, has demonstrated a dedication to the thorough exploration of a cave or a group of caves. The Award consists of a certificate and a cash award based on the income earned by the Lew Bicking Fund.

7. James G. Mitchell Award—The James G. Mitchell Award is given for the best scientific paper presented at any of the sessions of the annual convention by a member (or members) of the Society 25 years old or younger who has applied or been recommended for the Award. The Award consists of a certificate and a cash award based on the income earned by the James G. Mitchell Fund.

8. Conservation Award—The Conservation Award recognizes an internal organization which has demonstrated an outstanding dedication to the cause of cave conservation based on nominations from internal organizations or members. The Award consists of $50 from the Save-the-Caves Fund. The names of internal organizations that re-

ceive the Award are inscribed on a plaque which hangs at the NSS Office.

9. Peter M. Hauer Spelean History Award—The Peter M. Hauer Spelean History Award is given to an individual or group involved in an outstanding spelean history research project. The Award consists of a certificate and a cash award based on the income earned by the Peter M. Hauer Fund.

10. Certificate of Appreciation—Certificates of Appreciation are awarded by the President of the Society to people or organizations that have, by specific action, furthered the goals of the Society. The award consists of a certificate signed by the President.

By-Laws Committee

This committee is responsible for reviewing proposed NSS by-law changes and assuring that they are consistent with the Articles of Incorporation and with the Constitution of the Society.

Educational Opportunities Committee

The committee provides the membership of the Society with updated lists of education opportunities in both the scientific and sporting aspects of caving. The committee can provide information about courses and institutions with speleological emphasis.

Federal Agencies Liaison Committee

The purpose of this committee is to inform federal agencies of the information and talent resources of the Society and to otherwise facilitate an exchange of information and cooperation between federal agencies and the NSS committees and members.

Cave Vandalism Deterrence Reward Commission

The commission administers the Scoeity's program to provide cash rewards (up to $500) to those responsible for providing information leading to a confiction of persons who violate cave laws.

Finance Commission

The Finance Commission assists the Executive Committee in preparing the annual budget of the Society, advises the Secretary-Treasurer on the management of Society funds and the accounting techniques used, and serves the watchdog role of overseeing the Secretary-Treasurer.

Fund Raising Committee

The Fund Raising Committee coordinates national fund raising efforts. It is responsible for soliciting grants and donations to the Society's restricted funds (Save-the-Caves Fund, Cave Acquisition Fund, Cave Preserve Fund, NSS Support Fund, Ralph W. Stone Research Award Fund, Lew Bicking Award Fund, James G. Mitchell Award Fund, Peter M. Hauer Award Fund, Cave Safety & Techniques Research Fund), as well as unrestricted gifts to the Society. It also conducts specific fund raising efforts such as those currently authorized to purchase the Pettibone Falls Cave and to build an addition to the NSS office in Huntsville, Alabama. The committee will assist members and groups who wish to raise funds for a specific project (e.g., purchase a specific cave) on behalf of the Society. Contact the committee chairman for further information.

International Secretary

The International Secretary serves as coordinator between the NSS and the International Union of Speleology (UIS) and other individual and organization foreign contacts.

Legal Committee

The purpose of the Legal Committee is to serve as legal advisors to the Executive Committee and Board of Governors. The committee is not expected to serve as legal counsel to the Society. If you are an attorney and interested in helping, please contact the NSS President.

BOG Arrangements Committee

The committee solicits bids for the spring and fall Board of Governors meetings. Once the BOG has chosen the meeting location, the committee acts as a liaison with the meeting hosts to assure the necessary meeting arrangements are made. If your group would like to host a BOG meeting, contact the chairman of this committee.

Nominating Committee

This committee is responsible for conducting the annual NSS election and nominating candidates for NSS officers. Each fall, the Nominating Committee solicits potential candidates for NSS Director. It publishes its slate of candidates in time to allow additional candidates to be placed on the ballot by a petition signed by 25 voting NSS members. The committee publishes candidate information and distributes ballots to every voting NSS member in the spring. After counting the ballots returned, it announces election results. The elected directors assume office at the annual convention and serve a three-year term. Each year, four directors are elected.

While directors are elected by the general membership, NSS officers are elected by the directors. The Nominating Committee solicits potential candidates and develops a slate of candidates for each office. Additional candidates can be nominated by any director. The President and Secretary-Treasurer are elected at the spring Board of Governors meeting, while the Administrative and Executive Vice-Presidents are elected at the annual convention. Any member who desires to be a director or officer of the NSS should contact the Nominating Committee.

Public Relations Committee

In view of changing attitudes within the NSS with regard to public relations, this committee is becoming an increasingly important one. The committee is responsible for looking out for the NSS image in the news media.

The NSS News and grotto newsletters reach mainly cavers and subscribers. The Public Relations Committee can bring the personalities and philosophies of the Society to a wider audience. The committee is staffed by volunteer professional public relations specialists who use information to develop news releases and send them to appropriate hometown newsletters and other media.

Show Cave Liaison

Show caves serve an important need to the general public and can very well be an integral part of a program to conserve and preserve caves, cave fauna, and bats. Show cave development is one way to protect and preserve a cave. Even though the cave is modified somewhat through development, a carefully thought-out program of development does only minimal damage while preservation and protection from vandalism are virtually guaranteed.

It is important that the NSS maintain dialogue with show cave owners and operators. The Show Caves Liaison Committee is the NSS representative in this area. The committee visits show caves and talks with owners. They attend conventions of the National Caves Association (a show cave owners group). The committee also contacts appropriate government officials and agencies with regard to government-operated caves.

The committee has also tried to make NSS members aware of the bad reputation that cavers have with some cave owners, often caused by a few cavers who have been rude and disrespectful to show cave owners in the past. All cavers can be of service to this committee, to the NSS, and to themselves by acting like tourists when visiting caves and by demonstrating genuine interest in caves and the problems of their management.

Youth Groups Liaison

Scout troops, 4-H and Y clubs, high school science classes, and church groups sometimes include caving in their programs throughout the school year or as part of summer camp activities. Some of these groups have skilled cavers working with them, but others are led by adults with little or no caving knowledge, skills or experience. Public relations have suffered from the increased cave traffic, bad manners, and poor safety records of some of these groups.

The Youth Groups Liaison Committee's goal is to promote better cooperation between skilled cavers and youth-group cavers in the interest of safe, more responsible caving. The committee is working on establishing contact with youth organization on the national level. A set of caving guidelines has been developed for youth groups use and is available from the NSS office.

DEPARTMENT OF THE EXECUTIVE VICE-PRESIDENT

Under this department are also a number of committee, project groups, and task forces, each of which has been organized to handle a specific area of interest or study.

American Caving Accidents

This publication is a compendium of caving accidents in the United States during the previous year(s), complete with analyses and commentary. To date, the Society has published issues which cover cave accidents during the time period 1967 through 1981. The publication depends upon the reporting of all caving accidents to the editor of American Caving Accidents. This includes accidents involving non-affiliated cavers, as well as those affiliated in one way or another with NSS. The more reports on caving accidents that can be gathered in a given year, the more meaning American Caving Accidents for that year has in studying the reasons for, and possible means of having prevented, those accidents.

American Caving Accidents is dedicated to analyzing the causes and cures so as to make caving safer. The publication is an education in itself, for it draws awareness to the reasons for caving accidents, and can thus serve as an excellent training aid for safer caving activity.

Cave Files Committee

The NSS Cave Files are an archive of cave data. The files contain data on thousands of caves throughout the the world, although the emphasis is on caves of North America.

The purpose of the NSS Cave Files Committee is to:

- ☐ Solicit and store cave data.
- ☐ Maintain the NSS Cave Files.
- ☐ Serve as a center for information on the location of other cave data repositories.
- ☐ Establish a duplicate copy of NSS Cave File data to be separately located and encourage other cave data repositories to take similar precautions to protect the data from loss.
- ☐ Serve as a repository for original and duplicate cave data subject to such restrictions as may be mutually agreed with the donor.

NSS Cave Files data is available to NSS members and internal organizations on request, within reasonable limits and consistent with good conservation practice. The Cave Files Committee has a series of working rules to protect the files as much as is humanly possible from illegitimate and unscrupulous use. Cave data donated to the files is released only in accordance with any restrictions placed upon its release by the donor. Members can help the cause of the NSS Cave Files by feeding information on caves into the files, and by helping to protect the files data through using discretion in how and to whom they distribute data obtained from the files.

Caver Training Committee

The Caver Training Committee recently published *Caving Basics*, a manual intended for new cavers who have joined the Society. However, the manual should find appeal among all cavers, regardless of their caving experience.

Caving Information Series

This committee is responsible for publishing a series of inexpensive pamphlets on specific topics related to caves. They provide the detailed information cavers, grottos, or cave owners can use.

About the best way to describe these non-periodicals is to call them an open-file portfolio. It is a convenient and readily accessible source of practical information on the various and sundry subjects pertaining to caving. It is not complete and never can be, since it is an ever-expanding file of individual papers which deal in specifics on specialized subjects. It will appeal not only to the neophyte but to the knowledgeable caver as well. Since articles in such a file can become antiquated with time, the CIS committee rewrites and updates these papers.

The publications in the series are available to members at low cost from the NSS Office.

Conservation Committee

The primary purpose of the Conservation Committee is to coordinate and encourage the cave conservation efforts of the Society. The committee serves this purpose by acting as a communications network for the Society's various efforts. The committee also maintains liaison with individuals, corporations and state and federal agencies outside the Society, and coordinates Society conservation efforts with other conservation organizations having mutual goals.

Much of the efforts are handled by Conservation Divisions, Subcommittees, Joint Committees, and Conservation Task Forces (CTFs). In general, the divisions, joint committees, and subcommittees deal with national and regional issues of a continuing nature while the CTFs are chartered to work on local problems of specific interest to the CTF's members. In addition, the Conservation Committee tries to give support to local and regional conservation efforts when requested to do so by grottos, regions, and individual Society members. The committee also responds to requests for cave conservation information from outside the Society.

To provide financial support for conservation projects, Conservation Grants (up to $100 each) and Conservation Research Grants (up to $500) are available from the Conservation Committee. The Conservation Committee may also make interest-free loans to CTFs to facilitate carrying out their purpose and goals. The Conservation Committee receives monetary support through donations to the NSS Support Fund and the Save-the-Caves Fund, in addition to its regular budget.

If your group might benefit from being a CTF, write the Conservation Committee chairman. Any group of NSS members or an existing internal organization of the Society may be designated a Conservation Task Force of the NSS if the cave conservation purposes of the group and the Society would be aided thereby. CTF designation is made by the Conservation Committee chairman subject to BOG confirmation. This designation is in addition to any title the group selects for itself. For example, a group working toward the protection of XYZ Cave might call itself "Protectors of XYZ Cave—A Cave Conservation Task Force of the National Speleogological Society."

One of the main vehicles for communication of Society conservation efforts and encouraging further conservation projects is a regular monthly conservation column in the *NSS News*. Members and groups who are conducting a conservation-project are encouraged to send the Conservation Committee chairman a note describing the project so that the information can be used as a basis for further NSS News columns. By publicizing your accomplishments, others may be encouraged to try similar projects. Further, such publicity helps to establish a conservation ethic among the readership.

The following lists the various Divisions, Joint Committees, and Subcommittees of the Conservation Committee. It also briefly summarizes the primary goals or functions of each of the most active Conservation Task Forces. The Conservation Committee is organized into three Divisions (Information & Education, Fauna & Habital Protection, Management) which are permanent functioning units of the committee. The Information & Education Division includes the following subdivisions: NSS Membership, NSS Internal Organization, Private Cave Owners, Federal Agencies and Conservation Organizations. The Conservation Committee has also established two joint committees with the NSS Cave Conservation & Management Section. The joint committees are Newsletter and Grants & Awards. They were jointly established to carry out perpetual functions which if carried out separately would lead to duplication of effort. In addition the Conservation Committee has eight project subcommittees (Federal Legislation, State Legislation, Conservation Brochures, Federal Regulations, Resource Directory, Cave Owner's Guidebook, Conservation Photo and Cave Restoration Guidebook) to complete a specific task to a set time table.

Conservation Task Forces

1. Mammoth Cave System CTF—To coordinate Society efforts at Mammoth Cave National Park. The CTF has been active in assisting MCNP to establish a wild caving program and maintaining a liaison between cavers and the Park Service.

2. Pettibone Karst CTF—To save Pettibone Falls Cave and the surrounding karst. The CTF has been very active in attempting to get the current owners to sell the area to the Society.

3. Mt. St Helens CTF—To preserve Ape Cave and the surrounding Cave Basalt area on Mt. St. Helens. The CTF has successfully negotiated with the Forest Service to recognize the caves as a significant feature, to protect them from the mudflows caused by the eruption and to manage them appropriately.

4. Klamath Mountains CTF—To explore, map and classify the caves of the Klamath Mountain area of California. The CTF continues to map Bigfoot Cave and assist the Forest Service to inventory and manage their cave resources.

5. Sierra Caves CTF—To produce an inventory and management plan for Hummel's Cave in California. The CTF provided the Forest Service a draft management plan.

6. Mineral King CTF—To attempt to prevent a planned ski development for Mineral King, site of the highest known marble caves in California.

7. Peacock Springs CTF—To save Peacock Springs and the surrounding karst. The CTF was chartered in December of 1981. The owner of Peacock Springs Cave has offered to sell it, and the CTF is attempting to raise funds.

8. White River Plateau—To attempt to prevent limestone quarrying operations in an area on the White River Plateau, Colorado, which contains many caves. The CTF continues to be successful in preventing permits to allow quarrying.

9. Lost River CTF—To protect the Lost River Karst area of Indiana from a flood control project. Thus far, the CTF has been successful.

10. Germany Valley CTF—To protect caves and bat colonies of the Germany Valley of West Virginia, including Hell Hole and the Sinks of Gandy. These caves are threatened by quarrying operations and excessive visitation. Hell Hole has been declared a critical habit for an endangered bat species, and U.S. Fish and Wildlife has fenced the entrance.

11. Endangered Cave Fauna CTF—To work with state and federal agencies and within the scientific community for protection of threatened and endangered species and critical habitats. The CTF participated in publishing the CIS issue *Guide for Biological Collecting in Caves.*

Equipment Committee

The Equipment Committee is responsible for the scientific and other types of equipment owned by the Society. Requests to borrow the scientific instruments must be approved by the Research Advisory Committee.

History of the Society Committee

The Society has had a rich past since its formation in 1941. The purpose of the committee is to record and publish that past. Those interested in contributing old pictures and documents or helping to write the text should contact the committee chairman.

Landowner Relations Committee

One of the most ticklish problems with which cavers are faced is relations with the landowner. It is a fact of life that we cave on other peoples' property or on public land controlled by a government agency. Many grottos and regions have a landowner relations chairman who works toward keeping caves open in the grotto's locality. The national Landowner Relations Committee seeks to sensitize the membership by publishing regular articles in the NSS News on landowner/caver incidents, with an analysis of how the problem could have been prevented. If you know of an incident, drop the committee chairman a note describing it or contribute an article for the *NSS News*. A free brochure on landowner relations is available to members through the NSS office.

Members Manual

This committee's sole responsibility is to produce a manual for members. It is hoped that given a better understanding of NSS, individuals will become more active in the Society's programs while helping the Society to become more viable in the caving community.

National Cave Rescue Commission

The National Cave Rescue Commission is a volunteer group developed to coordinate cave rescue resources throughout the United States. NCRC itself is a communication network through which to locate the actual rescue workers and equipment. Most NCRC cavers do perform rescues, but not as part of NCRC; rather, as members of their local rescue squads, civil defense units or cave rescue groups.

Besides its main function as a communications network, NCRC has other goals:

1. Maintaining good working relationships with other rescue-oriented organizations, government agencies and sources of specialized equipment and services (e.g., the Air Force Rescue Coordination Center and the Mine Safety and Health Administration).

2. Maintaining current files of possibly useful equipment or services which can be obtained through the above sources.

3. Developing and maintaining a limited supply of certain equipment such as modified Neill-Robertson litters and rescue pulleys in key locations throughout the country.

4. Encouraging international cooperation by developing contacts with cave rescuers and evaluating existing equipment and techniques.

5. Increasing the number and proficiency of cave rescuers across the United States by sponsoring training sessions, seminars and workshops and by encouraging other cave rescue organizations to conduct such educational programs.

6. Encouraging international cooperation by developing contacts with cave rescuers and responsible agencies in other countries, by pre-planning with these groups where U.S. involvement in rescues is anticipated, and by inviting participation of cave rescuers from other countries in NCRC seminars.

NCRC is headed by a national coordinator and is divided into eight regional networks, each with a regional coordinator, selected by a board of all coordinators on the basis of recommendations from cavers and cave rescue groups in the region. In addition to the national and regional coordinators, the NCRC staff includes two specialists, a Cave Diving Officer and a Medical Officer, who provide advice and keep track of medical and dive rescue equipment and personnel. As do most rescue organizations, NCRC also depends on many volunteers with no official position, whose special knowledge, talents, or contacts make the network more effective.

A cave rescue is usually initiated by a companion or relative of the victim. NCRC may get involved at almost any stage thereafter, though many rescues can and should be performed with no NCRC assistance. Since very few non-caver agencies yet know of NCRC's existence and function, the Commission will probably be alerted by cavers. If the victim's companion or relative should contact the NCRC initially, the coordinator would get as much information as possible and then contact the closest or best equipped local cave rescue group. Time could be saved by making the initial phone call to the local group rather than NCRC. The leader of the local group, after deciding on and initiating the proper course of action, should call the regional coordinator to report that the group is going on a rescue and will be unavailable until further notice, and to request help in getting backup rescuers, specialized equipment, etc., if necessary. The regional coordinator may or may not advise the national coordinator at this point, depending on the scope of the rescue.

NSS Bulletin

The *NSS Bulletin* is a scientific journal on cave and karst research. It is the most widely circulated speleological journal in the world and is frequently used as a reference source for the writing of scientific papers. It is sent automatically to every member (except family members). Its primary goal is to contribute to the science of speleology and other specialties (social science, exploration, conservation, computer applications) associated with caves and caving.

The *NSS Bulletin* staff will assist any NSS member in preparing a manuscript for publication, whether or not that member is familiar with journal writing skills. The *NSS Bulletin* is open to anyone with new ideas and information of permanent value. This includes sport cavers as well as speleologists. If you are interested in contributing an article, contact the appropriate specialist on the Board of Editors.

NSS News

The *NSS News* is the Society's monthly news and feature magazine. It is sent automatically to every member. Within its pages, members can keep track of the numerous regional meetings held by caver groups around the country. Members are kept informed of Society business and affairs. In addition, there are articles of national interest concerning safety and equipment, conservation, regional news, and exploration, The position of News Editor is a part-time paid position. The job demands more time than a volunteer editor can reasonably be expected to give. You are invited to contribute articles, calendar announcements and other items of interest to cavers to this Society-wide publication.

Research Advisory Committee

This committee coordinates the research activity of the Society. It administers research grants given by the Society, chooses the winner of the annual Ralph W. Stone Research Award, designates official NSS Projects and Study Groups, provides advice to members concerning scientific efforts, answers speleological inquiries from others, and reviews proposals for research in Society-owned caves.

Projects of the NSS

Projects concerning caves, instigated by members of internal organizations of the Society, may be designated a Project of the NSS by the Research Advisory Committee (subject to BOG confirmation); if the proposer of the project would be advantageously served or recognized by such a designation. Current NSS Projects are:

1. Below Idaho: The Finite Wilderness—To create a 30-minute Public Television Program aimed at accomplishing the following goals: (1) to reexamine the human needs and social values served by wilderness preservation; (2) to present a simplified explanation of the 1964 Wilderness Act; (3) to establish a clear and easy-to-follow political and ethical justification for preserving wilderness; and (4) to investigate some of the problems facing the people charged with managing wilderness resources. By using caves as a vehicle to help hold the attention of a general viewing audience, this project will also endeavor to achieve the following: (1) promote an interest in protection and preservation of caves as a wilderness resource by creating an appreciation of caves; (2) to promote safe and sane caving practices; (3) to examine unique problems encountered in efforts to protect and manage wild caves in Idaho and (4) to further establish that caves are indeed a wilderness resource worthy of protection and management under the 1964 Wilderness Act.

2. The Bermuda Caves Project—To identify and document Bermuda's cave resources relating to history, marine and terrestrial zoology, botany, paleontology, geology and hydrology.

3. Caves and Karst Hydrology of the Lower Glen Rose Formation, Kendall and Comal Counties, Texas—To locate, survey and describe any significant caves in the region, to conduct a hydrologic study of the Lower Glen Rose Formation and relate it to cave development and to make landowners, land users and cavers aware of the caves and the need to protect them.

4. The Huautla Project—To explore and scientifically study the caves of the Huautla Plateau in Mexico, to develop new technology for the exploration of deep cave systems, to investigate the effects long duration underground camping on expedition participants and to publish a comprehensive report in book form. Several articles have been published to date.

5. Crystal Cave Dynamics Project—To study Crystal Cave, as ice cave in Lava Beds National Monument.

6. A Guidebook to Miner's Carbide Lamps Project—To develop and publish a book on the history and technology of carbide lamps.

7. Kingston Saltpeter Cave Research Project—To study the usage of Kingston Saltpeter Cave from the very earliest times to the present, to include a reconstruction of the environment and physiographic setting of the cave and surrounding area during the Pleistocene; the use of the cave by early American Indians; the mining of saltpeter during the Civil War; and the 20th century commercialization and recreational use of the cave.

8. The McFails Cave Cinematography Project—To produce a film and slide series on McFails Cave. The film is expected to be used to show the beauty and need for preservation of wild caves.

9. Oregon Caves Research Project—To produce a detailed map of Oregon Caves and carry out other research efforts.

10. Organ Cave System Project—To complete a survey of the Organ Cave system in Greenbrier County, West Virginia, and perform other research. Efforts continue on surveying, field checking draft maps and performing hydrological studies.

11. Stratigraphic Distribution of Caves in Northern Arkansas—To define the aerial distribution of caves in Arkansas, study measured section of Ordovician and Mississipian carbonates and relate this to cavern distribution, and relate cave distribution to local geology and speleologenesis.

Safety and Techniques Committee

The goal of the NSS Safety and Techniques Committee is "Safety through Knowledge," knowledge of safe techniques, safe equipment and awareness of the mistakes which cause most caving accidents. Activities of the Safety and Techniques Committee include caver educations, equipment study and testing, and development of safe caving techniques. Recent developments and educational material are communicated to the NSS membership through the committee's column in the *NSS News* and through workshops and seminars at NSS Conventions. While the emphasis is frequently on vertical caving, the area in which most serious accidents occur, the committee works on any safety and equipment problem. The Safety and Techniques Committee is current studying the effects of abrasion on various types and brands of rope.

Special Publications Committee

The committee arranges for the publication of works not published by other committees. Examples of the works for which the Special Publications Committee has been responsible are the *American Caving Accidents, Cave Gating, Cave Minerals, Caving Basics* and *Pursiking. The Jewel Cave Adventure* was published by Zephyrus Press through committee arrangements. This committee will review manuscripts and related materials for Society members and attempt to facilitate publication of those works which it feels will further the goals of the NSS and its members.

Speleo Digest

This annual was first published in 1956 by the Pittsburgh Grotto, which continued to publish it through 1966. The NSS has published the subsequent editions. A *Speleo Digest* contains selected material from local grotto newsletters of a given year, selected by a volunteer editor on the basis of its entertainment or permanent reference value. The *Speleo Digest* is thus a potpourri of information on American caves and caving. Nowhere else in America can one find such a comprehensive collection of written work on the activities of American cavers. *Speleo Digest* thus serves the important function of preserving, in a readily accessible form, much information from local newsletters that would otherwise be nearly impossible for cavers to locate in the future. *Speleo Digests* are not sold to the general public.

Study Groups of the NSS

A group of members or an existing internal organization of the Society may be designated a Study Group of the NSS if the scientific or technical purpose of the group would be advantageously served or recognized by such a designation. A Study Group is not an internal organization of the NSS, and the designation is in addition to the title of the internal organization with which it may be associated. It is intended that a Study Group be concerned with more general and longer-range efforts than encompassed under NSS Projects. Designation of a Study Group is made by the BOS. Current NSS Study Groups are:

1. Calaveras Speleological Survey—To encourage and coordinate cave studies within the Calaveras Limestone Formation of Central California.

2. Vermont Speleological Survey—To investigate, survey and describe geologically the caves of Vermont.

3. Virginia Speleological Survey—To compile new data on Virginia caves. VSS has published two books and is currently developing a third. It will be distributed to cavers and will include lists of the longest, deepest, wettest, etc., caves in Virginia.

4. Western Kentucky Speleological Survey—To study and describe the caves and karst features of western Kentucky. WKSS publishes annual reports describing their current work.

5. West Virginia Speleological Survey—To compile new data on caves in West Virginia and to publish bulletins concerning speleological research in the state. Seven bulletins have been published to date and two more (Salt Petre Caves of West Virginia and Caves of the Eastern Panhandle) will soon be published. Twelve more bulletins currently are being developed, and all should be published by 1985.

DEPARTMENT OF THE ADMINISTRATIVE VICE-PRESIDENT

Here again, the Department of the Administrative Vice-President is broken down into various committees to aid in their functions. Each is given a brief description in the following pages.

Audio-Visual Aids Library

The Audio-Visual Aids Library has a large collection of prepared slide shows, a great many of which include audio tape, which are available to NSS members and internal organizations. A nominal fee is charged to defray the cost of postage and maintenance. While many of these shows would appeal primarily to cavers, some are suitable for presentation to groups outside the caving community. A few are especially suited for youth groups and beginning cavers and can be used in training programs. These shows are produced by NSS members or groups and are donated to the library for the use and enjoyment of members. New shows are added regularly. The committee will be happy to assist members desiring to order slide shows or to produce new ones.

Bookstore

The Bookstore, housed at the NSS Office in Huntsville, offers publications and symbolic devices for sale, many at a substantial discount to members. The Bookstore has grown over the years and now offers some 75 different publications. An effort is made to provide commercial books relating to caves as soon as they are on the market. The various books published by the Society may also be purchased from the Bookstore. Still in stock are some back issues of the *NSS News* and *Bulletin*, many containing articles by cavers who have become famous in spelean history, and cave articles which will never become outdated. Orders for books or requests for the list of available titles should be directed to the NSS Office in Huntsville, Alabama.

Cartographic Salon

Another popular event at Conventions is the cartographic salon. Any NSS member may submit his best cave maps in accordance with the instructions in the announcement of the annual competition in the *NSS News*. A map exhibit is set up at the convention site. Judging takes place during the convention week based on art and technical merit and cartographic presentation. Awards are given in various categories based on size of map and mode of drafting.

The salon serves the following purposes: (1) to foster an interest in cave cartography as a technical skill and as an art form; (2) to allow an exchange of techniques, ideas and styles among surveyors and draftsmen; (3) to recognize cartographic excellence, and (4) to create an interesting convention display that depicts recent or active exploration and mapping projects worldwide.

Cave Ballad Contest Committee

Each year, the results of the annual cave ballad contest have been presented at the annual convention. These ballads (a reflection of our caving motivation and frustrations, our comedies and growing histories, our relative self-image, and the strength and quality of a caver's spirit) have long been a tradition of the underground culture. Records and tapes are produced incorporating some of the best songs submitted in the contest. The Cave Ballad Contest Committee exists to encourage cavers to express themselves through this art form in as imaginative and professional way as possible.

Congress of Grottos Chairman

The officers of the Congress of Grottos are elected by representatives of the Society's internal organizations and groups of unaffiliated members who attend the congress. Each congress elects the COG officers which will serve through the end of the following congress. The COG officers are responsible for soliciting issues for COG discussion, preparing the congress agenda, chairing the meeting, and submitting the results of votes to the BOG.

Convention

The highlight of the year for many cavers is the NSS Convention. Held during the summer months, it is located in various parts of the country from year to year. As might be expected, an affair of such national scope reflects the many facets of cavers and their activities.

First, the NSS Convention provides the opportunity for cavers to make presentations on their activities in the exploration, conservation, management, and study of caves. Separate sessions are devoted to such fields as geology, biology, hydrology, social science, exploration, conservation, management and spelean history. In addition, there are geology and biology field trips, conducted to explain local physical conditions or problems which are open to speleologist and sport caver alike.

Seminars and workshops command a prominent place of importance. Here, members can discuss, in a relaxed group participation atmosphere, a wide range of subjects pertaining to virtually every aspect of caving. Some of the subjects that have been addressed by recent workshops are photography, grotto publications, use of computers in caving, problems of large cave projects, conservation, stream tracing, vertical techniques, and public relations.

The Board of Governors meets in its summer session, and the Congress of Grottos holds its annual meeting. NSS sections and committees hold their meetings and luncheons at this time. It is a time to review the year's accomplishments and next year's plans. The Photo Salon displays the best in cave photography and the Cartographic Salon presents the finer side of cave mapping. Results from the annual Cave Ballad Contest are presented, the participants submitting original and adapted words and lyrics, some of which will be sung around cavers' campfires for years to come. There is a vertical contest, sponsored by the Vertical Section, with numerous classes according to one's age, the type of climbing rig, and the distance climbed.

Just for fun, there are the nightly campground gatherings, often complete with bonfires and song. There is the Howdy Party, a massive Convention get-together for fellowship, good food and drink, and often with the entertainment of a band, and maybe even a surprise or two.

The Speleolympics is an obstacle course contest which simulates conditions in a subterranean environment, but seldom resembles anything even remotely approaching a cave, except possibly for the mud and water. It is a race against the clock, and despite the scrapes and bruises, both participant and spectator have fun at this sometimes bizarre spectacle.

The NSS Bookstore and other vendors sell equipment, publications and virtually anything a caver might need. The Convention week is climaxed by the banquet and the presentation of Society awards of every imaginable sort. After a presentation by the evening's guest speaker, everyone heads home with fond memories of the completed Convention and anticipation for next summer's happening.

Internal Organizations Committee

The Internal Organizations Committee is responsible for chartering new internal organizations and maintaining a current file on all active IOs. Each IO must submit an annual report to the committee; those which fail to do so are eventually deactivated.

The committee currently publishes the Administrative Memo, sent to all active internal organizations, NSS Leaders, and individual subscribers upon payment of $5 a year. A subscriber also receives monthly office mailings, including membership updates and BOG minutes. The Memo contains news primarily of interest to IOs and NSS leaders, with suggestions of ways to promote the NSS through IO activities. The committee also has published memoranda on such topics as *The Small Grotto (How to Start and Maintain)* and *A Guide for Local Newsletters* and annually circulates a list of publications by all active IOs. Additionally, the committee works closely with the Congress of Grottos and serves as the credentials committee for that body.

Library

The NSS Library is located at the NSS office in Huntsville, Alabama. The library regularly receives issues of over 200 different speleological periodicals a year from NSS internal organizations and other cave organizations around the world. A budget is provided for the library to purchase significant cave-related publications. The library answers requests for information and loans out some items of which there are duplicates. Generally, however, the library serves as a significant research facility for anyone interested in the study of caves.

Membership Committee

In an effort to become more effective in its conservation efforts and scientific endeavors and to increase its membership services, the NSS is actively attempting to increase its membership. The Society's infrastructure can easily increase the level of services to an increased membership. A larger membership means more people to work on new or existing conservation and scientific projects, and more money available to fund such projects. NSS members comprise only a small part of the caving population. There is a large group of unaffiliated cavers which the NSS can solicit as members. These cavers can help further the goals of the NSS and can benefit from the Society's safety and conservation messages.

Due to the fragile and non-renewable nature of caves, the NSS restricts its membership recruiting to those who are already interested in caves rather than trying to recruit people who have never been caving before. The Board of Governors voted unanimously that, "The effort to increase membership shall be directed primarily to persons that are already involved in some activity relevant to speleology." However, membership is open to all who want to join.

The NSS in recent years has experienced a real explosion in terms of new members coming into the Society. It took roughly the first 18 years of the Society's existence to reach the point of having a total of 10,000 members, both past and currently active. It has taken only the last ten years for the Society to add another 10,000 members. The total active membership, however, has shown only a gradual increase. Fewer than half of the members who joined the NSS in the last ten years have remained active in the Society. The real challenge for the Membership Committee and the Society as a whole is not in recruiting new members, but in keeping the interest of those cavers who do join.

NSS Museum of Speleology

The purpose of the museum is to collect objects directly related to caving and speleology. Museum materials include vertical gear, lighting devices and other equipment, grotto and region patches, and other memorabilia. The Museum of Speleology provides a significant display of historic

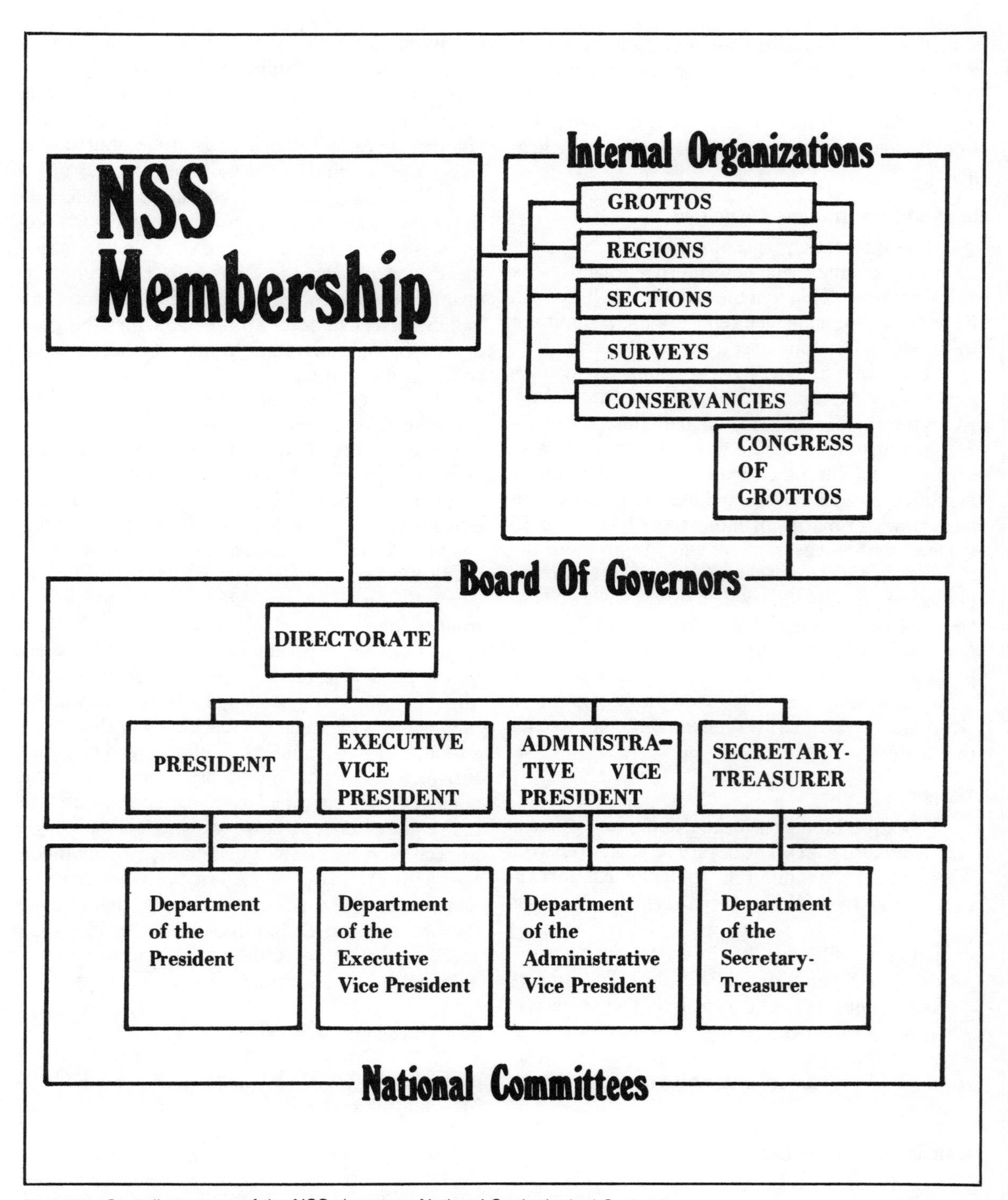

Fig. 10-1. Overall structure of the NSS. (courtesy National Speleological Society)

materials at the International Congress of Speleology and plans to have similar exhibits at future conventions and other functions.

NSS Office

All national organizations with a significant number of members need a central headquarters. The NSS purchased its office in Huntsville, Alabama in 1972. The NSS Office is often the first segment of this complex and far-flung Society with which new members have direct business dealings. At the Office, new and renewal membership applications are processed. The Office answers all Society mail, directing inquiries to the proper channel. All problems with regard to membership are directed to the NSS Office. The Office keeps available an assortment of brochures and pamphlets, types and mails the monthly membership list updated, changes addresses, runs the NSS Bookstore, and keeps the key to Shelta Cave, the entrance to which is on the adjacent property.

The volunteer office manager is responsible for keeping the wheels of operation well-oiled. Under the supervision of the office manager are the only two full-time paid employees in the NSS. They perform all of the above-mentioned duties plus many more.

Photo Archives

The Photo Archives Committee was established in an attempt to provide a depository for slides and photographs of caves and the people who have been concerned with them. Identification of caves and cavers in photographs now in the Archives continues. Members are requested to consider contributing photographs in their possession (fully identified) to the Archives.

Photo Salon

One of the highlights of any NSS Convention is the Photo Salon. Any NSS member is invited to submit his best photos to the Photo Salon Committee following announcement of the annual competition in the *NSS News*. Judging takes place prior to the NSS Convention, and awards are given in several categories on the basis of technical excellence, aesthetic beauty, originality, and artistic merit. The categories judged are black and white prints, color prints, color slides, and movies.

Winning prints are exhibited in a gallery set up at the Convention for conventioneers to browse through at their leisure. The slides are presented during an evening assembly late in the week of the Convention, where awards for all categories are formally announced and background information about individual photos is provided. This program has consistently been the most popular Convention event.

Program and Activities Committee

The Program and Activities Committee has the responsibility of seeking sponsors for NSS Conventions, and aiding organizers in convention planning. The committee serves as one of the points of contact for those wishing to use the NSS name in a list of sponsors for an activity or event. The committee also brings to the BOG applications for official NSS status as field trips or expeditions.

A visit to a cave or a cave area with NSS members in charge may be designated an "NSS Field Trip" if the purposes of the trip should be advantageously served by such official designation. A major trip to a cave or a cave area, usually on a large scale or to a distant location may be designated an "NSS Expedition."

Field trips and expeditions organized and conducted by members, grottos, regions or other internal organizations of the Society may be called (for example) "The ABC Grotto (NSS) Field Trip" or "An Expedition of the ABC Grotto of the NSS."

DEPARTMENT OF THE SECRETARY-TREASURER

The Department of the Secretary-Treasurer maintains the financial records or "books" for the Society. The books consist of a journal and a ledger. Each financial transaction or check is recorded in the journal as it occurs. The NSS also files an annual information tax return with the Internal Revenue Service. The current NSS fiscal year runs from May 1 to April 30 of the following year. This department is made up of three committees.

Cave Ownership and Management Committee

The Cave Ownership and Management Committee was created for the purposes of studying and making recommendations to the BOG on proposals to acquire and manage new cave properties, and to oversee and report to the BOG on the management of currently owned caves. This committee is also expected to develop a consistent and workable long-term program for the ownership and management of caves in general. Thus, the work of this committee will not only apply to the management of NSS-owned caves, but to those caves owned by persons outside the Society as well. The committee will endeavor to answer any questions on cave management and encourages groups or individuals to ask for assistance relating to cave management.

McFail's Cave Committee

McFail's Cave, with five miles of mapped passage, is the largest cave northeast of Virginia and West Virginia; it is owned by the NSS. An excellent physical description of McFail's is found in the March and May 1979 issues of the *NSS News*. Thanks to the efforts of NRO work parties and a few dedicated individuals, the cave's parking lot and lawn are kept mowed and a registration box gives information to the caver as well as records the cavers in the cave. A half-mile nature trail with signs take visitors through the karst area between the two entrances. Both entrances are gated and locked, and a camping area has been provided.

Besides the BOG appointed chairman, the committee consists of one member from each grotto in the NRO and a local NSS groundskeeper. All access to McFail's is handled by written permission through the chairman. Any non-NSS group must petition the BOG for admittance. The committee's functions include reviewing any cases where denial of access has been questioned and debating special projects, such as the effect that recent pushes in exploration might have on the cave's ecology or landowner relations. The chairman will only grant permission to qualified owners who will wear wet suits, and each has his or her own vertical gear. If a group wishes to use the new entrance (Hall's Hole), they must have at least one member who is familiar with the old entrance (Ack's Shack) in case of rising water.

Throughout NSS ownership, research has been conducted in and around McFail's Cave Preserve. Currently, the emphasis has swung away from geology and toward biology and meteorology. The July 1981 issue of the NSS News has an article on the recent bat survey conducted in cooperation with the New York State Department of Environmental Conservation, Endangered Species Unit. There is an ongoing project to record the air and water temperature and the stream height in the cave.

Exploration in the McFail's system has led to exciting discoveries and opened other possibilities. The new entrance, Hall's Hole, was first opened in 1978, and it has been most welcome in terms of comfort, time savings, safety, and rescue. Rock engineers have recently scaled two sizable domes in the cave, and cave divers pushed the "terminal" siphon to more passage. There are four other small caves on the Society's property and numerous sinkholes, so opportunities are there.

The McFail's Committee has had excellent relations in the community, as evidenced by the nearly total lack of litter or vandalism on the property. The committee has assisted the local geology classes in surface tours in a karst cave as well.

Shelta Cave Committee

It happens that, hardly by coincidence, the NSS Office at Huntsville, Alabama, sits atop a cave. Bill Torode, chairman of the committee, describes his duties thusly:

"Since I live across the street, it is easy to keep an eye on the property to keep out motorcycles and run off kids cutting down trees or trying to get into the cave. I walk around the property once a week and pick up trash, keep the weeds trimmed down and the outside neat, keep the gate repaired, and fix the wooden ladder from time to time as the wood rots and has to be replaced. Occasionally I take groups into the cave, both NSS members and non-members." Others also contribute their time and efforts to protect the cave. Shelta Cave is open to NSS members.

John Guilday Cave Preserve Management Committee

The most recent cave purchase by the Society occurred in March 1983 when the NSS bought the Trout Rock Caves (Hamilton, Trout and New Trout caves) near Franklin, WV. Together, those caves contain over seven miles of passage and significant paleontological sites. The caves are open to the public. Those interested in visiting them should contact the committee chairman.

Kingston Saltpeter Cave Committee

In 1983 the Society entered into an agreement with the Georgia Kraft Corporation and the Felburn Foundation to lease the Kingston Saltpeter Cave property and manage the caves. The committee is currently actively involved in conducting archeological studies and performing a variety of cave management activities.

Anyone who is even vaguely interested in cave exploring should be a member of the National Speleological Society. The information provided from this fine organization is an invaluable aid and a constant resource for all spelunkers. Perhaps, you're not interested in cave diving at this time, but if you are, the NSS has information on this pursuit. Climbing techniques, alpine caves, organization, and many other aspects discussed or alluded to in this book are actively practiced by many NSS members who regularly contribute to NSS information bulletins, booklets and other printed material. The NSS is a friend to all cave explorers who are willing to abide by their rules and regulations, which mainly address safety and conservation. Without the NSS, amateur spelunking could not have progressed to its current healthy state.

Glossary

acrophobia: An unusually strong fear of being in high places. In regard to cave exploring, high places are relative. That is, you are almost always beneath the surface of the earth, but you can still be confronted with sheer drops which put your current location at a much higher point than the floor below. Acrophobia should not be confused with the fear of falling, which is a basic human instinct. Those who suffer from acrophobia may panic while near the top of tall buildings, even though they are completely enclosed and protected from falling. Acrophobes typically are frightened on airplane flights, in any high place, and sometimes even at the thought of being in a high place. For this reason, acrophobes should not attempt to explore caves where substantial drops may be encountered.

aisle: Usually a narrow passageway or a narrow passage within another passageway. Large passageways are sometimes divided into two or more sections by obstructions which emanate from the floor but do not continue all the way to the ceiling. Each of these sections is then known as an aisle.

alabaster: A commercial term for gypsum, which is often mined from caves. (see *gypsum*)

anchor: Any point to which a rope or rope ladder is attached for descent into a fissure or opening in a cave. The anchor must be of adequate mechanical strength to support the rigging and the weight of the spelunker or spelunkers and all equipment during the descent and ascent. Spelunkers must carefully choose natural anchors in caves. These often take the form of large stalagmites, boulders and other such projections. Most safe anchors contain smooth surfaces that will not cut the rope when weight is applied. When a number of spelunkers hold onto a rope while another descends to scout out new passageways, this is known as a ***human anchor***.

anthodite: A commercial word used by Skyline Caverns in Front Royal, Virginia to describe highly unusual calcite formations that look like delicate flowers. The term itself is a Greek derivative, meaning cave flower. Anthodites partially resemble aragonite and other helictite formations. Their extreme rarity has never been fully explained.

aragonite: A usually snow-white calcium carbonate that has formed under pressure. This type of formation many take may different forms. Sometimes it appears as a delicate flowstone or a white stalactite or stalagmite, as well as in highly irregular patterns. Aragonite basically consists of the same materials as other common but less attractive cave formations. Its white appearance is attributed to the method by which it is formed.

Archeozoic Era: In the earth's development, the Archeozoic Era dates back to over a billion years ago and is the earliest era of geological history. The earliest known rocks date back to this period and are probably the only remnants to survive in modern times.

back door: A slang term among spelunkers which often relates to a second entrance or exit from a cave, usually one that is smaller or less obvious than the main entrance or *front door*. The term may also apply to the last person (the one bringing up the rear) in a spelunking expedition; here, the *front door* is the group leader. This type of exploring arrangement is often used when inexperienced persons are taken through a wild cave by more experienced individuals, two of whom are at the front and rear of the party. The "back door" has the responsibility of making certain no members stray from the group.

backswim: A slang term used among spelunkers to describe an exploring method where water is involved. This technique relates to an exploring situation where a pool or stream of water is narrowly separated from the cave ceiling. It then becomes necessary for the spelunker to enter the water back first, keeping the nose and mouth above the surface to take advantage of the narrow, waterless breathing space. This technique can be highly dangerous when practiced by inexperienced spelunkers or when it must be used for long distances and in unknown passageways.

backup light source: In cave exploring, any device that will generate light but which is not *normally* used; an emergency lighting source should the main source of illumination fail. Typically, a carbide lamp will serve as the main light source, while backup lighting consists of a flashlight and candles. Alternately, the backup source may be another carbide lamp, along with a flashlight and/or candles. Cave exploring safety rules require that each spelunker carry at least two backup sources of light.

bacon: Commercial and/or slang term for a type of flowstone formation consisting of parallel bands of alternate dark and light colors. Bacon formations are often advertised as highlights of commercial caverns. This formation is often translucent and can occur when a section of flowstone (or some other formation) is broken. The resulting formation resembles sidemeat or commercially packaged bacon strips.

balcony: Any horizontal projection emanating in an underground room or passageway and roughly resembling a balcony on an aboveground structure. Balconies may be of any size from a few inches to several feet in width and depth. A true balcony is usually large enough to comfortably support a spelunker above the cave floor.

base camp: The main meeting point during a spelunking expedition; it may be located aboveground, but is more often found underground during overnight expeditions. The base camp normally contains sleeping arrangements, food, and supplies, along with other gear which is not normally carried on explorations of short duration. When underground, the base camp is often established at a central point within a large cavern and forms the hub of an exploring wheel with passageways going off in all directions from the hub. Base camps are useful for large expeditions where many people are involved, small groups of which may strike out on discrete explorations. All report back to the base camp within a certain prearranged time period. One person is always left at the base camp in order to offer assistance if needed and prepare camp meals and other needs.

bat: A nocturnal flying mammal with four limbs which have evolved to form wings. These furry creatures can be found in varying numbers in most caves throughout the world. There are many different species, and all belong to the order *Chiroptera*. In the United States, commonly seen bats include the little brown bat, long-eared bat, hog-nosed bat, and a few others. The little brown bat is most often seen east of the Mississippi. Spelunkers may encounter only a few bats in a particular cave, and several million in other large underground caverns. While bats have been known to carry rabies, incidents involving spelunkers and these animals are quite rare. Most bats tend to sleep peacefully on the cave's ceiling, provided that spelunkers do not deliberately disturb them.

bat roost: A room or rooms in a cave in which bats seem to congregate, roosting during the daytime hours. Actually, anyplace a bat chooses to spend the daylight hours could be termed a bat roost, but true roosts are typified by large quantities of bats grouped together. During the nighttime hours, the bats exit the roost and can sometimes be seen in quantities of 100,000 or more.

bedrock: The solid layer of rock which lies under all loose surface materials. This is a solid base which may be formed by sedimentary, igneous, or metamorphic rock.

belay: A safety rope which is used to protect the climber during an underground exploration. *Belaying* is a technique used to attach and use the safety rope. Normally the spelunker attaches the rope around himself and above the waist most often with a bowline knot. Once the knot is tied, the rope may be adjusted to fit comfortably under the armpits of the climber. Again, this is a safety rope to prevent a fall should the main rope break or the climber becomes dislodged from it. The *belayer* is the anchor man at the top of the drop who plays out rope to the person making the descent as it is needed. The belayer's body is well braced and he is instantly ready to apply full pressure should an incident occur.

blind fish: Any of a number of small fish with functionless eyes. Hundreds of thousands of years of living in the total darkness of underground caves has negated the need for eyes, so evolution has slowly phased out this portion of the sense organs and apparently developed in these animals a strengthening of their remaining senses. Cave fish typically are albino, containing no skin pigment. Due to the relative scarcity of food in underground streams when compared to those on the earth's surface, these mutant animals are quite small, requiring an equally small food supply to survive.

box work: A pattern of thin plates of calcite protruding slightly from limestone walls and ceilings. This is a reticulated pattern resembling closely spaced boxes or parallelograms. Box work is not especially rare in caves, nor overly plentiful either. This interesting formation may take several different forms.

breakdown: Also known as a *cave-in*, a ituation where the cave ceiling collapses. Breakdowns are often caused by construction work in an area over the cave, especially where digging and blasting are involved. In many breakdowns, materials from the surface of the earth actually fall down into the cave passageways. Fortunately for spelunkers, natural breakdowns are extremely rare.

calcium carbide (CaC_2): A dark gray crystalline material often shattered into gravel-sized bits for use in carbide lamps. When calcium carbide comes in contact with water, acetylene gas is produced. In carbide lamps, water is contained in one chamber and is gravity-fed to a lower chamber containing the calcium carbide. As the water drips on the carbide, the acetylene gas is formed and travels upward through a tube, finally exiting at the lamp nozzle. Acetylene gas is highly flammable but is not especially dangerous in low-pressure applications such as the carbide lamp. When the escaping gas is ignited at the carbide lamp nozzle, a small flame is produced that, with the aid of a reflector, provides light for spelunkers. *Caution:* When calcium carbide (for lamp refills) is taken into a cave, it must be securely packed in an absolutely watertight container. Due to the dampness of underground caverns, the humidity in the atmosphere alone can cause acetylene gas to be emitted from exposed carbide.

candle: A backup light source carried for emergency use by many spelunkers. Underground safety requires three different sources of light, which are most often the carbide lamp (main source), a flashlight, and finally candles. While almost any type of large candle may be used as an emergency light source, most spelunkers prefer a cylindrical type about three inches in diameter. Candles make very poor light sources, but they are relatively dependable in that even when they are dropped in water, the portion of the wick surrounded by wax will remain relatively dry. It is then a matter of carving away the top section to expose the dry portion of wick. While candles cast very poor light, they can be used when the other two sources have failed. Caves are often drafty, and one is constantly plagued by the flame being blown out. This requires regular relighting, but many have been saved by including a candle or two as part of their gear.

candle test: Some caves are low in oxygen content and contain high amounts of carbon dioxide. This is especially true of some smaller caves having only one very small entrance. Using a candle, these types of situations can often be detected before trouble develops. The candle test simply involves lighting any candle and carrying it with you into the cave. If the candle flickers and then goes out and cannot be relit to burn for any length of time, this is a sign of low oxygen content, and the cave should be exited immediately. This is a fairly reliable test and should be performed any time a question exists as to the oxygen safety of a cave. Carbide lamps, too, will burn with a very low flame in low-oxygen environments, but the candle will usually not burn at all.

carabiner: A tool used in underground climbing and rope work. It consists of an oblong ring with an adjustable link, allowing it to snap to the eye of a piton in order to hold a freely running rope. The rope slides through the carabiner during this type of operation. The carabiner allows for the path of a rope to be controlled and, with the piton, for pressure to be applied. Carabiners are sometimes used in descending and ascending drops which are composed of several vertical and horizontal levels. This prevents the rope from coming in contact with sharp rock ledges which can cause abrasions.

carbide: See *calcium carbide.*

carbide infection: A condition which results from getting expended carbide dust into the eyes, nose, mouth, and especially an open wound. This is a very painful infection which can occur when spent carbide is not disposed of properly or where a spelunker with a cut or open wound about the area of the hands or wrists changes carbide in a less-than-professional manner, spilling it on his body. Any time carbide comes in contact with an opening in the skin, the exposed area should be washed thoroughly, treated with an antiseptic, and then covered. Carbide infections should be treated by a qualified physician.

carbonation: A term derived from carbon dioxide and means to impregnate with this gas. The carbonation process is the cause of such cave formations as stalactites, stalagmites, columns, etc. (see *carbon dioxide*)

carbon dioxide: A colorless gas (CO_2) that dissolves in water, forming carbonic acid. It is formed by the combustion and decomposition of organic substances and is a large part of what is exhaled during the human breathing process. Carbon dioxide does not support combustion, and the high presence of this gas usually indicates a lack of oxygen. Some caves have high levels of carbon dioxide (and low levels of oxygen) and are highly dangerous to cave explorers unaided by breathing apparatus. (see *candle test*)

cave: A *natural* underground room and/or series of passageways and other such openings formed by water rushing through underground rock strata. To be truly classified as a cave, such networks must have at least one or more openings to the earth's surface.

cave art: Normally describes the prehistoric drawings on the underground walls of caves that were used by early man. These often depict hunting scenes and date back to the days when humans first appeared on the face of the earth.

cave channel: In an underground cavern, the prime passageway. This will mark the gravity flow of the underground stream which initially formed the channel. Caves often have many subchannels which are normally smaller passageways that run, generally, perpendicular to the main channel. These were formed by offshoots of the main stream. The cave channel is the one most often traversed by spelunkers during the beginning stages of an exploration. Once this large passageway has been fully explored, the subchannels are then entered and sometimes mapped by experienced parties. The cave channel will always run from and to the entrance of the cave.

cave club: An organization of spelunkers which meets regularly for the purpose of discussing and exploring caves, proper techniques, safety, etc. Some are loose organizations of individuals; others are quite structured and may involve many members. Many cave clubs are affiliated with the National Speleological Society and adhere to their practices and regulations. In the United States, there are approximately 100 such organizations registered with NSS.

cave coral: A formation that closely resembles natural coral, but which is composed of calcium carbonate. This may go by several other names and is neither rare nor abundant in most American caves.

cave diving: The exploration of partially or totally submerged underground passageways. This is a highly specialized form of spelunking and must never be attempted by anyone who has not had the proper training in *underground* diving techniques. Most divers always have the option of surfacing, but the underground diver may not have an air-containing surface immediately overhead. The actual exit may require traveling through several long passageways. Cave diving is most often done using standard scuba equipment, along with safety lines and underwater lighting. The NSS offers a cavern diving course which serves to introduce the student to the cave environment under the supervision of an experienced cave diving instructor. The course lasts one weekend and includes lectures on diving philosophy, the cavern environment, safety procedures, plus three actual cavern diving sessions. Upon satisfactory completion of the course, the student is awarded a cavern diver card by the NSS Cave Diving Division.

cave fauna: The animal life which is found in any cave. This normally precludes some animals that may temporarily take refuge in a cave but do not normally live there. For example, foxes will often make a den in the opening of a small cave, but are not considered to be cave fauna, since this type of domicile is usually temporary and may not be repeated year after year. Bats are true cave fauna, although they do not live continuously in the underground environment. However, bats can always be found in caves and use the underground chambers as their main residence.

cave fill: Any material which completely or partially blocks a passageway. This is usually a natural material such as mud, gravel, or stone and which was not a natural part of the cave's actual formation. Cave fill is often silt which remained when the rushing waters that formed the cave exited to lower levels. It may also be mud that was brought into the formed passageways from the surface by small streams which formed during the rainy season and flowed for a short period of time every year. Over the years, such buildups can be tremendous and whole passageways and even rooms are almost completely filled.

cave flora: Plant life found living in a cave. More flora is usually found at or near the cave entrance due to the presence of some obstructed light at this area. Deep underground, one will often find some forms of algae. Cave flora does not include the calcium carbonate formations which are sometimes referred to as *cave flowers*.

cave flower: Any stone formation which closely resembles a living plant. There are no true natural flowers which grow in the depths of a cave. (see *anthodite*)

cavern: Technically, a cave and a cavern are one and the same. However, the term cavern is often a relative one: *Cave,* then, normally describes a smaller network of underground passageways, while *cavern* describes a much larger system which typically contains large rooms, underground streams, and many sub-networks branching from the main passage. A cavern, then, is a large cave.

cave system: All of the rooms, passageways, channels and any other aspect of a particular cave. The system starts at the opening and ends at the farthest point or points underground. Some cave systems are very small, containing only a single passageway; others are extremely complex and are made up of mile upon mile of winding passageways, rooms, underground streams, etc.

ceiling: The roof or highest section of the underground passageway, room, or channel which can be seen or accessed. Actually, most caves have many different ceilings, as they are divided off into many different rooms and passageways. Technically, the term ceiling should be applied only to the roof of an actual cave room. However, the term is most often used to describe the highest point in any particular portion of a cave system.

Cenozoic Era: A period in geological history that extends from the start of the Tertiary Era to today. The Cenozoic Era is marked by rapid evolution of flora and fauna, such as mammals and birds, but by little change in invertebrates. In cave exploring terminology, the Cenozoic Era especially applies to the system of rocks which were formed during this period. The beginning of the Cenozoic Era dates back some 70 million years and is the era we are currently in. It was immediately preceded by the Mesozoic Era, which extended from the period of about 200 million years ago to the beginning of the Cenozoic Era.

chasm: A deep cut in the earth. In cave exploring, a chasm is often an irregular gorge encountered along a passageway. It is normally marked by a sizable drop and differs from a pit in that it is typically long and fairly narrow. Chasms may be explored or may be bypassed by simply traveling over them using ropes, ladders, etc.

chicken loop: A slang term used by cave explorers and mountaineers alike. It is the loop which is formed in the bottom end of a rope through which the foot is placed. The explorer hangs onto the rope with his hands, and with foot securely in the chicken loop, he is lowered by fellow spelunkers to the bottom of a drop or raised from the bottom to the top.

chimney: A steep vertical passageway going upward from the cave floor. Actually, the term chimney may depend on whether you're on the bottom or the top of the drop. At the top a chimney is known as a *well*, but at the bottom, the term chimney is quite appropriate. This, of course, assumes that there are passageways at both the bottom and the top of the drop. A chimney is typically less than five feet in diameter and may often spiral to an upper level of a cavern system.

chimney walk: A method of climbing up a chimney using no special tools. Since chimneys are usually fairly narrow, the spelunker places his back against one wall, his feet against the other, and then, applying pressure, pushes himself upward. This is a fairly easy process to master, but it is quite tiring and should not be attempted without safety lines, especially when large drops are encountered.

chin cup strap: While most persons recommend that hard hats used for cave exploring purposes not be equipped with any type of securing strap whatsoever, some do recommend a chin cup strap. Using this arrangement, the strap does not fit behind the chin but rather centrally terminates in a soft cup that is molded to fit the chin tip. This prevents possible strangulation (when compared to the under-the-chin strap) should a helmet become lodged in a crevice. (The author does not recommend any type of helmet strap for cave exploring purposes.)

chin strap: A device fitted to a helmet for the purpose of holding it firmly in place. It fits under the chin and can cause strangulation in cave exploring work should the helmet be lodged in a crevice during a fall. (see *chin cup strap*)

claustrophobia: An abnormal fear of being in closed or narrow spaces. While claustrophobia is a psychological condition which can be treated at least to some degree, a caving environment will still have adverse effects on most persons with even *slight* susceptibility to this disorder. People who do not normally suffer from claustrophobia will sometimes experience slight twinges when spending a long time in an underground environment, especially when fatigue sets in. These twinges do not come close to approaching true claustrophobic conditions which are often evidenced by nausea, extreme perspiration, paranoia, and actual panic. Persons with claustrophobia should never attempt cave exploring unless their condition is completely under control. Even then, extreme caution is mandatory, since a cave environment is the ultimate in complete enclosure.

cleft: Any irregular opening in the cave floor, wall, or ceiling. In most instances, the term is used to describe a small and often shallow fissure but is a relative term that can even describe large chasms.

column: A formation created when a stalactite and stalagmite grow to adequate length to fuse at their respective tips. When this occurs, the water that initially made the two formations ceases to drip and begins to flow from the cave ceiling to the cave floor. After many hundreds of thousands of years, the point where the two formations initially connected becomes covered with calcium carbonate, and the tapering effect of the stalactite and stalagmite is converted into a single formation of fairly uniform cross-sectional width from cave floor to cave ceiling. This type of formation is sometimes called a *stalacto-stalagmite*.

commercial cave: Usually a large underground cavern system that has been modified to allow access to the general public. The conversion process usually requires the removal of cave fill in order to make all passageways high enough to provide easy access without stooping. Additionally, passageway floors are specially treated to remove obstructions and make them fairly level. Guard rails are placed around any dangerous fissures or openings and an artificial lighting system is installed throughout the passageways and rooms to make up the tour. There are many commercial caves throughout the United States, some of which are open year round. Typically, visitors to the cave do not require special clothing and are led through by a tour guide. Prices vary for commercial cavern visits, depending upon the length of the tour, the size of the cavern, etc.

commercial tour: The trip through the passageways of a commercial cave. (see *commercial cave*)

conduit: The course of an underground stream through a passageway. These passages are completely filled with water at all times and require cave diving equipment for exploration. A conduit may also be any small or cross-passageway which links two larger rooms or passages.

connection: A general term used in cave exploring that often describes the exact point at which a passageway opens into what can be classed as a room. It may also be a passageway between two larger passageways or rooms in a cavern system.

controlled climate: Due to the fact that most caves are separated from the earth's surface and hence its weather conditions by many layers of stone, the internal environment of the cave exists in a controlled climate. Cave temperatures and relative humidities do not change substantially with the changing of surface weather conditions. If a normal cave temperature in the summer is 52° Fahrenheit, then this same temperature will be maintained in the winter months when the temperature dips below freezing on the outside. Caves are not affected by sunlight, wind, and most other weather conditions with the exception of rainfall, which can sometimes swell subterranean streams to overflowing. The controlled climate of a cave places fewer adaptation requirements on cave flora and fauna. Animals, for example, do not have to physically adjust to changing weather conditions, as they do aboveground, and flora does not cycle with the change of seasons, nor day/night sequences.

corkscrew: Usually a small tunnel or passageway formed by spiraling waters. This effect leaves smooth ridges along the length of the passage resembling the appearance of the kitchen utensil used for removing corks from bottles. When corkscrew passageways are more or less vertical, they are sometimes known as chimneys. (see *chimney*)

corridor: A passageway that is relatively straight and connects two other passageways, sections, or cave rooms. The corridor is a hallway between two other points in the cave system. (see *conduit; connection*)

coveralls: The standard attire of the cave explorer. It is a onepiece suit which normally has no belts and covers the body completely from ankles to neck. In underground exploration, coveralls are typically worn over more traditional clothing and serve to prevent the body from becoming snagged on sharp ledges and projections by providing a uniform clothing surface. Since the coveralls are of one-piece design, cave dirt, mud, etc., is partially blocked from reaching the major portion of the skins surface.

CPR: Cardiopulmonary resuscitation, a life-saving procedure often practiced on accident victims in deep trauma. CPR courses are taught by the Red Cross and other organizations in most communities and are highly recommended for anyone interested in spelunking. This procedure has saved many lives, both aboveground and below.

crawl: A slang term for a narrow, low passageway or area of a cave which can only be accessed by crawling. The crawl technique requires that the spelunker lay flat on the floor, face up or face down, and then pull himself along using feet, hands and knees. This is a tedious process, and due to the roughness of most cave floors and obstructions, can lead to assorted bruises and bumps when proper protection is not provided in the form of clothing.

crayfish: A freshwater crustacean closely resembling a lobster but much smaller in size. Blind crayfish are sometimes found in the depths of subterranean pools and streams. While closely resembling their surface counterparts, these creatures are usually much smaller, and have lost most or all of their pigmentation, making them a dull white. The discovery of any aquatic life which has evolved underground is a bit of a rarity in the caving world, so such animals should never be removed from their natural environment. Blind crayfish have their roots in normal surface dwelling crayfish that were washed underground and took up existence there, eventually evolving into the pygmy, eyeless forms seen today.

crevice: A fissure in the cave floor, ceiling or wall. It is a relative term and usually describes an opening that is somewhat bigger than a cleft but smaller than a chasm. (see *cleft; chasm*)

crossover: A passageway that may take a spelunker over a fissure or chasm or around some other difficult area of the cave. It is often associated with a passageway which lies above a room or other passageway, but may also refer to an interconnecting tunnel between two rooms.

crystal cave: A cave that contains an abundance of translucent crystalline formations. Instead of the dull browns and grays that color most cave walls and formations, crystal caves contain forms of these same materials that reflect light. Others are pure white and seem to glow under the illumination of the carbide lamp. Crystal Cave is also a commercial cavern discovered by the late Floyd Collins, a professional cave explorer in the early part of this century. It is surrounded by Mammoth Cave National Park near Cave City, Kentucky, and consists of a network of galleries and corridors estimated to cover more than sixty miles. Many areas of this cave are replete with crystal formations. Collins lost his life in this cave when he became trapped while trying to discover a new entrance alone.

cul-de-sac: A passageway that terminates in a dead end. The ending point is often slightly larger than the rest of the passage and may even be a small room. Cul-de-sacs have only one entrance and are often known as dead-end passages.

curtain: A calcite formation running vertically from the ceiling toward the floor which is marked with clearly defined ridges formed by the union of a row of stalactites. Flowstone formations are often present on the curtain's surface. When curtain formations are very large, they are sometimes called drapes or draperies.

dead cave: One in which water has ceased to trickle down through the ceiling. This process causes all formations to stop growing. In time, the tiny particles which make up these formations will crumble away. Sometimes caves will go through a dead period and then due to geological changes, water will flow again and the formation process will continue. (compare with *live cave*)

deep cave: Caves are initially formed by flowing water. In some instances, the water will exit the underground chambers at some point aboveground. In other situations it will drop to a lower level, form another cave, and then drop even further. A deep cave consists of two or more main levels, the lower of which may often be completely filled with water. Deep cave is also a relative term describing underground passages which lie a great depth below the surface of the earth. Deep caves are usually much older than the shallow caves which are small and were formed on only one level.

developed cave: See *commercial cave*.

domepit: A domepit is a vertical shaft which was formed by water descending along a joint plane. It is often larger at the top than at the bottom and may consist of various lengths, widths, and heights.

drapery: See *curtain*.

driftstone: Sections of cave formations that have shifted in relationship to the walls and ceilings. This shift is brought on by hidden fissures, often in the floor of a cave, which cause this section to move minute distances while the walls and ceiling remain in place.

electric lamp: A source of illumination used in cave exploring that operates on electrical power. A flashlight is the most simplistic example of an electric lamp. More sophisticated forms consist of a bulb and reflector followed by a lens housed in a separate compartment from the batteries and mounted to the hard hat. A battery package is worn around the waist (inside the coveralls) and is attached to the lamp with a flexible cable. Some spelunkers use electric lamps as a main source of illumination. However, they are more frequently associated with backup and emergency light sources. When the electric lamp is worn on the hat, the connecting cable to the waist-mounted battery pack is often subjected to snagging on projections and is therefore not as highly regarded as the self-contained carbide lamp.

entrance: Most often, the aboveground opening that leads into a cave. Some caves may have more than one entrance, but most contain only a single opening which is also the exit. The term entrance may also be applied to a point where an aboveground stream suddenly travels underground or to a subterranean passageway which leads to another section of the cave system.

entrapment: In cave exploring, a situation where a spelunker or exploring party is unable to exit the cave. This can be brought on by many different circumstances, but most often describes a condition where a breakdown has occurred, thus blocking the exit, or rising waters have filled the exit completely. Spelunkers have been entrapped because their light sources failed and they simply could not see to get out. Other reasons include rope or equipment failure, fatigue, etc.

erosion: The act of wearing away a surface by means of water action or by any other medium that creates friction. Most caves were originally formed through the erosion process. The friction of the flowing water removed the loose strata, enlarged fissures, etc., while the chemical reaction of the water with limestone caused further weakening. As these particles came away from the main rock surface, they were washed away by the flowing water.

excavation: The digging out of a cave or cave entrance for better access. Many commercial caverns have required extensive excavation to remove cave fill and other materials in order to enlarge the passageways for easy touring. Spelunkers sometimes dig through cave mud in hopes of uncovering another passageway or to remove obstructions. Any underground excavation must be done with extreme caution, because cave fill often holds large boulders and other objects in place. When the fill is removed, these obstructions can suddenly shift, creating a potentially dangerous situation.

expedition: A trip through a wild cave, as opposed to a tour through a commercial cavern. The expedition is made up of the spelunkers and all equipment. Some expeditions are completely planned regarding their route through the cave, while others are only planned to a certain point which will be followed by seeking out previously unknown portions of the subterranean passageways. The term expedition usually refers to a well-organized cave exploration where a great amount of planning has gone into the selection of spelunkers, equipment, etc.

filler cap: Usually a hinged arrangement located on the top portion of the carbide lamp which exposes the opening to the top chamber that is filled with water. The cap is removed to fill the lamp and then pressed back in place to prevent spillage during an expedition. Most filler caps contain a small hole in their centers to allow a small amount of water to escape when it is put under too much pressure by the gas generated within the lower section of the lamp. (see *carbide lamp*)

fissure: See *cleft.*

flashlight: See *electric lamp.*

floor: The area which lies between the walls and ceiling and is the main travel route for spelunkers. In the underground there are many different floors, as there are different rooms and passageways on different levels. The floor is the bottommost section attainable in any passageway or room on any one level of the cave, as opposed to balconies, for instance, which are extensions from the cave wall that can be walked over but lie above the true cave floor.

flowstone: A formation deposited on the walls, floor, and sometimes the ceiling of a cave by minerals contained in flowing water. Flowstone formations still show evidence of the flowing motion which originally created them. Wavelike surfaces can easily be seen. Flowstone is often relatively smooth, especially when a light covering of water still exists on its surface. It is quite common in most caves and one of the more attractive formations.

formation: In cave exploring, any mineral deposit that has been placed on the roof, floor, or walls of a subterranean passageway or room. Formations include stalactites, stalagmites, flowstone, helictites, etc.

gallery: In deep caves with two or more levels, all of one level. A gallery may also be an unusually large room in a cave system (in relation to other rooms) which contains a large number of formations.

gear: All the equipment which a spelunker takes underground during an exploration. In its simplest form, the spelunker's gear consists of a coverall, a hard hat, carbide lamp, and two emergency lighting sources. Complex explorations however, dictate complex gear such as tents, sleeping bags, scientific equipment, ropes, alpine tools, etc.

geology: The science dealing with the history of the earth and its life as recorded in rocks and minerals. Caves are especially useful in geological studies, as the rocks contained there have been relatively unaffected due to the controlled environment. Geology is closely aligned with speleology in many aspects. (see *speleology*)

grotto: A small cave or a small subsection of a larger cave. This is a somewhat relative term and uses a comparison of the terms cave and cavern. Using cave as the central reference, a grotto is smaller, while a cavern is much larger. A grotto is also an authorized group of the National Speleological Society.

ground water: Water that lies a short distance below the earth's surface and is composed of all moisture in the part of the ground which is fully saturated. Ground water may eventually seep down through the cave ceiling, where it is then referred to as *subterranean water*.

guano: In the speleological sense, bat excrement. This can accumulate in quantities of tons in caverns that contain large numbers of these creatures. Guano is highly flammable and has been mined at times from the interiors of caves for energy purposes. A guano cave is one which contains a large number of bats and thus large quantities of guano. These caves can be very dangerous for a number of reasons. First, the flame from the carbide lamp could conceivably ignite the guano. In some cases, this amount of flammable material could set up a tremendous explosion. Guano caves may also contain a large amount of methane gas, which is poisonous.

guide line: A small-diameter rope used by spelunkers to identify a proper route of exit. This especially applies to cave diving, where a network of water-filled passageways must be navigated. The first diver in locates the guide line, while following divers use it to mark their route. When it comes time to exit the passageways the line is simply followed in reverse order. Guide lines are required safety equipment for all underwater diving.

gypsym: A commonly found mineral ($CaSO_4 \cdot 2H_2O$). It contains of hydrous calcium sulfate. This mineral sometimes makes up entire caves, but more often, cave formations.

halazone tablet: A water purifier often used by spelunkers who intend to stay underground for long periods and who must drink water found in the cave streams. Halazone is a purifying agent and removes harmful impurities which can affect health.

hard hat: A piece of safety equipment mandatory for all cave exploring. The hat is often made from a tough fiber or plastic and is commonly seen in use around construction sites. Underground, the hat protects the spelunker's skull in case a rock should fall from some height. It also serves as an impact guard in low passageways where the head may bump the ceiling. Hard hats have saved many a spelunker from serious or even fatal injuries. They are rugged pieces of equipment that can stand a lot of abuse, but which should be replaced when any signs of serious damage exist. Hard hats should not be worn with chin straps; rather, the headband should be adjusted for a firm, comfortable fit. Most spelunkers equip their hard hats with mounting brackets for the attachment of a carbide lamp. When these are properly adjusted, the lamp beam will focus on a point directly in front of the wearer's eyes. This frees the hands for holding ropes, crawling through low passageways, etc.

headache: A term used in the overhead construction industry for many years as a warning to persons below that an object has been dropped. The same use is applied in cave exploring. When one spelunker has other members of his party directly below him when making a descent, if he should kick loose any stones or other debris, he will immediately call out "Headache!" Upon hearing this cry, the cavers below will seek immediate cover to protect themselves from the falling object.

helictite: An erratic lateral projection of calcium found in caves. Its appearance is similar to that of a bare twig, but it may also be seen in clusters of tiny needles, contorted needles, and even in a formation that resembles rime ice. Helictites are seen in many caves throughout the United States, and they loosely fall into the cave flower category in that they do vaguely resemble some surface plants and flowers.

hypothermia: A condition of abnormally low body temperature. The body is a remarkable machine that is quite efficient at regulating its temperature to a normal average of 98.6° Fahrenheit. When subjected to high external heat, perspiration forms to serve as a cooling process. However, when the body becomes too cool, there is no built-in regulator for this extreme. Clothing helps to conserve body heat, but when dampness occurs, the heat is quickly bled off. The caving environment is a cold and humid one. Temperatures average from the high 40s to the middle 50s. Such temperatures are not considered dangerous if the proper attire is worn. However, when a long stay in a cave is necessary, spelunkers must guard against continually wearing damp clothing. Persons can die of hypothermia when exposed to temperatures even in the high 50s for long periods of time, especially when injured or weak from hunger. Hypothermia robs one of strength and severely reduces reflexes and other mental functions. It is a major cause of caving incidents and accidents.

hypoxia: A condition caused by a deficiency of oxygen. This differs slightly from asphyxia, which is a condition of low oxygen and/or high carbon dioxide level. This latter is usually caused by the interruption of breathing, which brings on unconsciousness. Hypoxia is a condition which involves the body's tissues and can occur in situations where there is adequate breathable oxygen, providing that some physical condition exists whereby the oxygen does not reach the tissues of the body. This can be brought on by shallow breathing and a number of other things. It is sufficient to say that every spelunker should be aware of general oxygen levels in all caves explored. Some caves are high in carbon dioxide content and low in oxygen. This is especially true of small caves with relatively small openings to the outside. (see *candle test*)

ice cave: A cave which forms in a large glacier, iceberg, etc. When formed in icebergs, these caves are often partially or totally submerged. Such passageways are brought on by fissures and weaknesses in certain sections of the ice and through which warmer water is allowed to flow. A more common definition of *ice cave* is one in which ice forms and generally remains throughout most of the year. These are sometimes found at high altitudes or in extremely cold climates. Many have several large openings to the outside that allow a constant subfreezing breeze to blow through, thus dropping the overall temperature to below 0° Centigrade. Dripping water then forms icicles which look very similar to stalactites. In fact, many other formations are sometimes replicated in ice within a short period of time. This is a much speeded-up process when compared to the hundreds of thousands of years required to form most true cave formations.

igneous rock: Rock formed through heat and subsequent cooling and hardening of the molten materials. Igneous rocks are made from magma or lava that was originally contained deep within the earth's interior. Igneous rocks are formed very suddenly when lava spews from a volcano and then rapidly cools.

isopod: A small sessile-eyed crustacean with a body composed of seven different segments, each of which bears two legs. Isopods have been discovered living in caves. Their development follows the pattern of most other lifelong cave dwellers in that their eyes have degenerated and all body pigment has been removed by evolution. These tiny creatures are rarely seen by those not actively seeking their presence.

keyhole: An opening or passageway in a cave whose top section is substantially larger than its bottom, giving it the general shape of the keyhole opening found in many doors. Some keyhole passageways are large enough to permit standard traverse, while others require the spelunker to crawl to the top, larger portion and proceed from there. These types of openings are quite common in many caves, although one has to use a bit of imagination to detect a true "keyhole" shape.

kit: In spelunking, a caving kit normally consists of a small knapsack or body bag which contains a minimum of two emergency light sources, some high-caloric food such as candy bars, nuts, etc., spare carbide, and a few first aid items.

kneepad: Commonly available devices with a tough rubber surface, roughly rectangular, and molded to fit over the kneecap. Two side straps hold it in place. They are used by those who must subject their knees to a lot of abrasive surfaces. Spelunkers find these items very useful in protecting their knees while crawling through passageways. They are generally available at most hardware stores.

labyrinth: A maze of passageways and complex paths formed within a section of a cave. These passageways are often very tiny and extremely numerous. They often form near large underground pools or lakes, but may be found at other locations as well. Sometimes, large cave systems are called labyrinths because maps of the various passageways resemble highly complex mazes.

ladder: In spelunking usually a collapsible device made of ropes or metal cables, between which rope or metal rungs are installed. These ladders are quite useful in descending sheer vertical drops. They are attached at the top to an appropriate anchor and allowed to simply drop to the bottom. The spelunkers then descend and ascend, but not in the typical fashion used on a carpenter's ladder. One climbs up and down collapsible ladders by straddling one edge and locking the feet and hands into rungs on both sides. Safety lines should always be used when spelunkers are on collapsible ladders.

lamp gasket: A device that produces lighting by generating acetylene gas from carbide housed in the lower of two chambers. The top chamber contains water which is allowed to slowly drip onto the calcium carbide below. The gas is formed in this bottom chamber. Since it is necessary to regularly change the calcium carbide charge, the bottom chamber simply unscrews and is removed from the main lamp body. To assure a good leakproof seal when the bottom chamber is reinstalled, a rubber gasket is fitted at the back of the threads and should mate tightly with the main lamp portion. Since the seals are made of rubber, they eventually age and the seal will be broken. When this occurs, acetylene gas is not completely contained for diversion to the nozzle, and some of it leaks around the gasket. If the gas is ignited, the whole lamp seems to go up in flames. This situation can occur at almost any time; should it happen while its owner is descending a sheer drop, a serious accident could occur.

lamp tip: The nozzle of the carbide lamp. This is a tiny device inserted into the neck opening of the lamp. In its face it contains a tiny opening which allows a narrow jet of acetylene gas to escape. The gas is ignited at this point and the increased pressure due to the small aperture of the nozzle keeps all flame outside the lamp proper.

lead: In underground terminology, usually a small narrow passage which branches off from a main passageway. These are often evidenced by low ceilings which necessitate laborious crawling for traverse. (see *crawl*)

lily pad: A circular buildup of translucent minerals on the still surface of a subterranean pool. These occur from the continuous dripping of water containing minerals from a ceiling that often is only a few inches from the surface of the water. If the ceiling were much higher, the force of the water droplet striking the underground pool would disperse the minerals and no buildup would occur. When the droplets falls from only an inch or so away, the water surface is not greatly disturbed and the deposited minerals remain in place. Lily pads form horizontally (in parallel with the surface of the water), so they seem to grow in width rather than in height as do most other cave formations. Lily pads do not often take as long to form as other types of cave formations, but their growth rate is still measured in the thousands of years.

lily pond: Is a pool containing lily pads; also, a pool which contains stalagmite formations that rise from the bottom and then expand at the surface of the water, similar to water lilies. This type of configuration is somewhat unusual. The stalagmite formations originally formed before the pond was present and reached upward to a ceiling that was close overhead. At a later time, water washed away most of the ceiling material, leaving the blunt tops which were originally the bases of stalactites. The water remained at the base level, thus the lily pad effect.

limestone: A sedimentary rock whose composition is made up largely of calcium carbonate. The great majority of caves in the world are formed in this type of material.

live cave: One which is still forming, i.e., streams are still flowing and carving out future passageways. A live cave is also one whose formations continue to drip water. Each droplet deposits a microscopic amount of minerals which is evidenced in the appearance of growth after thousands of years. (see *dead cave*)

lockjaw: A common name for the infectious disease *tetanus* whose characteristics include muscle spasms, especially of the jaw, and eventually death if not treated. It is caused by the toxin of the Tetani bacillus which is usually introduced to the human body through a break in the skin. This bacillus is almost always present in cave mud and caving environments in general. For this reason, it is mandatory that everyone interested in cave exploring be immunized. Even then, all cuts or breaks in the skin should be treated with an antiseptic and bandaged immediately.

maze: A confusing, intricate network of passages in a cave. (see *labyrinth*)

medical kit: A small, watertight container containing simple medical supplies such as adhesive bandages, antiseptic, smelling salts, and other items to treat minor injuries. Large expeditions, especially those which plan to spend more than 24 hours underground, often carry highly sophisticated medical kits containing splints, braces, and even certain drugs which can be administered by trained personnel.

Mesozoic Era: The Mesozoic Era dates back approximately 200,000 years and is the period in the earth's history when dinosaurs and other such life forms existed. The Mesozoic Era lasted until about 70 million years ago, when the Cenozoic Era began (see *Cenozoic Era*). It was during this time that many caves were forming and was an era of great change within the earth's interior.

mooring: The act of securing caving gear or a spelunker with a safety line. Also, the safety line itself; the term is most often used thus. A mooring is a small, flexible rope which is often used to guide a spelunker across a gorge or to give him lateral control when suspended on a rope.

natural bridge: A formation which spans a fissure or chasm. It is formed when the sections of stone supporting it are stronger than surrounding sections and thus are not washed away by the flowing waters that initially carve out the cave. Some natural bridges are of miniature proportions, while others are large enough for one or more persons to safely walk on. Sometimes a natural bridge is formed when a large formation such as a stalactite, stalagmite or column breaks from the cave floor or ceiling and falls across a fissure or gorge. Natural bridges sometimes make convenient exploration paths through caves. However, some have been known to give away suddenly and should not be trusted in many instances.

nodule: A small bump, generally oval in shape and quite often smooth, that forms on a cave wall, ceiling, or floor. A nodule is often the beginning of a stalactite or stalagmite formation. Some resemble fried eggs.

NSS: The National Speleological Society, a nonprofit organization devoted to the study of caves and related phenomena. It was founded in 1941 and is chartered under the laws of the District of Columbia. The Society is associated with the American Association for the Advancement of Science. The NSS has provided leadership, guidance, and training for persons interested in cave exploring for over forty years. It is a highly respected organization and one which offers a myriad of resources and services for its members throughout the world. One of the main goals of the NSS is conservation of caves.

nursery: In bat caves, a large room housing a great number of pregnant female bats and those nursing young bats. Some nurseries may contain many thousands of bats, all hanging vertically from the ceiling. These sights are very rare and are more common in the western part of the United States, where large concentrations of bats in caves seem to be more numerous.

overhead chamber: A separate room which connects laterally to another passageway or room near its top. To reach an overhead chamber, it is necessary to scale the cave wall.

overnighter: A planned caving expedition that requires members to pitch camp and actually spend the normal sleeping hours underground. Overnighters require much more planning than a normal expedition, which would typically last less than five hours. Members of such an expedition must take special precautions regarding hypothermia because of their increased exposure time. Also, extra medical supplies, food, sleeping gear, and the like are needed.

paleontology: The science dealing with the development of life within past geological periods. Paleontology depends upon the examination of rocks and fossil remains to reconstruct what must have occurred millions of years ago.

Paleozoic Era: A period in the geological past which extends backward some 600 million years. The period ended about 200 million years ago. During the Paleozoic Era, fishes, amphibians, reptiles, and insects began to appear on the earth. Over this period, the earth changed in many different ways. There were periods where warm, humid climates brought about the growth of coal-making plants, great limestone deposits occurred, etc.

pillar: A column of rock usually extending from cave floor to ceiling. It may be similar in appearance to a column, but pillars are not classified as true formations. Whereas a column occurs when two formations (stalactite and stalagmite) eventually grow together, a pillar forms through the solution process—or, more accurately, through a lack of it. The area immediately surrounding a pillar was solid at one time millions of years ago. As water flowed through and over this mass of minerals, all were dissolved away with the exception of those which made up the pillar. The pillar, then, is composed of materials which are stronger or less soluble than those around it and is all that remains of a solid layer of minerals. Pillars do not grow in the same manner as stalactites and stalagmites. However, if water is dripping or flowing over the pillar surface, a layer of flowstone can develop after many thousands of years, and small stalactites and stalagmites can form at its top and base.

pit: A hole in the cave floor; the term is generally associated with one of fairly large diameter. A pit is also referred to as a cave well and is generally circular in diameter. (see *well*)

piton: An implement used by rock climbers consisting of a short, metallic stake containing an eye or a ring at one end. The point of the piton is blade-shaped and designed to be driven into a small crevice in a rock ledge or wall. The climber uses a piton hammer to drive the piton home. The climbing line is then attached and the piton depended upon to support the climber.

quartz: A mineral made from silica or sand. It is sometimes found in caves and consists of a silicon dioxide. The chemical designation is SiO_2. Quartz sometimes occurs in colorless form whereby the crystals are transparent. Alternately, it is seen as colored hexagonal crystals and also in great masses resembling rock candy.

quartzite: A type of sandstone. It is a compact, granular rock composed of quartz (or more accurately, quartz sand) and is derived from sandstone by metamorphism. (see *quartz*)

rappel: A technique used in rock climbing whereby the climber lowers himself by rope using the friction of the rope against and around his body to control the speed of descent. The rappel is a very basic climbing technique and is often used for descending sheer drops in caves where it is impossible to attain a hold on the side of the wall. Rapelling is used when free drops are made; that is, the climber is suspended by his rope in open space quite some distance from any wall. Using the rappel, the rope is passed under one thigh, across the body, and then over the opposite shoulder. This configuration provides the climber with leverage to easily control his rate of descent. When he wants to descend faster, he loosens his hand grip a bit, allowing teh rope to slide across his body. When it becomes necessary to slow his descent, he applies more grip and can even bring himself to a complete halt if so desired.

rigging: A term most often applied to climbing maneuvers in caves. The rigging consists of the climbing implements such as ropes, ladders, pitons, etc., and how they are arranged for the ascent or descent. Rigging may also refer to the gear the spelunker wears or carries into the cave. (see *gear*)

room: An enlarged portion of passageway within a cave that expands outward roughly at 90° angles from the passage. It may have an entrance and an exit, several, or only an entrance alone. In cave exploring, room is a relative term and is a comparison of the size of one area of passageway to the general size of all other passageways.

rope burn: A condition which occurs to the human skin, often but not limited to the hand area, which is caused by a climbing rope sliding across the skin at a great rate of speed so as to create heat due to friction. The result is a burn (from friction heat) and/or a relatively fine laceration. Rope burn is quite easy to get when climbing without gloves. When rappelling, rope burns can occur about the thighs and buttocks if proper rigging is not worn. Rope burn is usually not serious, although a severe burn garnered during a descent within a cavern may make it very difficult or impossible to make the return ascent. Proper gear and rigging during all climbing operations will go a long way to lessen the possibility of severe rope burn.

ropework: A term which applies to all of the procedures used to make ascents and descents by means of ropes and climbing accessories. A person's ropework is a description of how well he climbs and handles himself during climbing situations. Ropework may also refer to the rigging of ropes, pitons, safety lines, etc., made for a particular ascent or descent.

rubble heap: Generally a pile of rocks, broken formations, etc., which may block a passageway or simply be found in the middle of a subterranean room. A rubble heap can be caused by a ceiling collapse or, more often, by a large formation which has become dislodged from the cave ceiling and breaks into many pieces upon striking the floor. Sometimes large sections of the cave ceiling will give way (often due to aboveground construction work). While this collapse may not actually create another opening to the cave (near the collapse site), it can bring down tons of rubble. This is a problem when some caves are being prepared for commercialization. Due to the excavating (and sometimes blasting) which occurs, rubble heaps are formed in areas of the cave far distant from the actual work being performed. All rubble heaps must be removed for touring convenience, and naturally, all loose sections of the ceiling that have the potential to become rubble heaps must be removed.

safety rope: A secondary rope which is attached to the climber (often by a bowline) to serve as a backup should the first rope break or the climber loses his grip on it. The safety rope is secured to the climber, so it is not necessary for him to grip it should an accident occur. At the top of the drop, a human anchor is established, with this spelunker lowering the safety rope at the same rate as the climber's descent.

salamander: An amphibian closely resembling a lizard. Salamanders, however, do not have scales and are covered with a soft, moist skin much like that of a frog. Salamanders go through two stages of development. In the larval stage, they breathe by gills; in the adult stage, they lose their gills and breathe with tiny lungs. This is equivalent to the tadpole/adult stages of a common frog. Cave salamanders are quite an unusual find. These animals have lost most of their pigmentation, and their eyes have degenerated to small lumps which are completely useless as visual aids. Cave salamanders are smaller than their aboveground cousins, and some never go through an adult stage. Many cave salamanders never lose their gills and remain in the water all their adult lives, eating the microscopic animal and plant life brought in from the outside by rainwater.

sandstone: A sedimentary rock that usually consists of quartz sand held together by silica or calcium carbonate. Caves are sometimes formed in sandstone, but these are more often caused by wind erosion than by the process which has formed the vast networks of a more common limestone cave.

scouting party: A group of spelunkers exploring aboveground areas that are likely to contain caves for the purpose of finding cave openings for later underground exploration. Scouting parties can be quite useful for mapping out a day's spelunking activities in areas where cave openings are found. The scouting party may enter some caves in order to check out their potential for a later exploration. When this occurs, there should be at least three members and a full complement of spelunking gear must be carried underground, even if the stay is to be a brief one. Based upon the findings of the scouting party, the spelunkers may elect to visit one or more of the caves reported. Sometimes, when large groups take to the underground, a central point or camp is established and scouting groups of three or more spelunkers set out in different directions in order to locate passageways which might make interesting exploration for all. This is another form of scouting party, but regardless of what the party is trying to do, it should *always* consist of three or more members and follow the equipment and safety rules of a standard underground expedition.

scuba: Diving equipment used for exploring subterranean passageways that are completely filled with water. Scuba is an abbreviation for self-contained underwater breathing apparatus. This consists of a pressurized tank of air, a harness to hold it in place on the diver's back, and a regulator which keeps airflow at a more or less constant pressure, regardless of relative underwater pressure. Scuba equipment is self-contained and does not require an air line to be maintained between the diver and a compressor on the surface. This allows the cave diver to move freely without having to worry about the hose becoming torn on sharp rock ledges.

shallow cave: One where all passageways are contained on one main level. (see *deep cave*)

shield: A disk-shaped formation similar to a curtain and often occurring when two stalactite formations grow together. Shields are often covered by flowstone and are roughly rectangular in shape, resembling the shields used in ancient times for hand-to-hand combat.

sinkhole: A visible depression in the surface of the earth, often found near or over underground caverns. It is often representative of a collapse in the cave ceiling which has caused the upper layer of the earth's surface to move downward. Sinkholes can be very large, measuring several hundred feet across, or very small. The sinkhole does not usually result in a new opening to the cave, as surface debris quickly fills in any new entrance that might be formed by such a collapse. When geologists or spelunkers suspect a large cavern system in a certain area, they often look for the presence of sinkholes. If a natural opening cannot be found into the cave, a shaft is sometimes dug in or near a sinkhole in an attempt to create an artificial opening.

special equipment: Implements and apparatus necessary to perform highly specialized explorations (the carbide lamp, coveralls, ropes, etc., may be thought of as more or less standard equipment). Special equipment might consist of scientific test instruments, elaborate photographic devices, communications apparatus, etc. Cave diving equipment also falls under this specialized category. (see *scuba*)

speleology: The scientific study of caves and other subterranean phenomena. It is derived from the Latin word *speleum*, meaning cave. A *speleologist* is a scientist and should not be confused with a spelunker, a person who makes a hobby of exploring and studying caves.

speleophobia: The fear of caves. This is partially a slang term that describes other more recognized phobias, such as claustrophobia (fear of enclosed places), acrophobia (fear of heights), etc. All of these fears are felt by a true speleophobe.

speleothem: A term given to items which are taken from caves and caverns, such as formations, plants, animals, etc. The collecting of speleothems should be discouraged, since if practiced by many people, will quickly destroy the natural beauty of subterranean passages and chambers.

spelunker: A person who makes a hobby of exploring caves. Spelunking may be thought of as a sport, one which requires physical stamina and training. A spelunker should not be confused with a speleologist, since the latter is a person who scientifically studies caves, cave formations, etc. The word spelunker may be a derivative of the Latin word for cave, but a popular explanation says that amateur cavers got their name from the fact that when you drop a rock into an underground pool it goes "spelunk."

spelunker's tuxedo: A humorous slang expression for the drab coveralls and other standard attire of amateur cave explorers. (see *coveralls*)

stalactite: A common formation in most caves usually made from calcium carbonate deposits left on the cave ceiling by dripping water high in mineral content. Stalactities resemble icicles in that they are large at the base and drop vertically to a fine point. Some stalactites are very small, measuring less than an inch in length. Others may be twenty or thirty feet long and weight many tons. Stalactites always form from the ceiling downward or from a high spot on a wall to a lower spot. Often, a mirror image of the overhead stalactite forms below it on the cave flow. This is called a *stalagmite.*

stalactostalagmite: When a stalactite and stalagmite eventually grow together, a new formation is created which is a solid mass of minerals from cave floor to cave ceiling. As this formation continues to grow, it no longer resembles either of its two parent formations. Today this is most often called a column, but another term which is just as correct is stalactostalagmite. Some formations such as these are less than a foot high and were formed by the joining of two very tiny stalactite/stalagmite formations. Others may weigh several hundred tons and be quite massive in both width and height.

stalagmite: A calcium carbonate formation that begins at the floor of the cave and extends upward to a fine point. It is often the mirror image of the stalactite which lies directly overhead. Usually, a stalagmite can only be formed from water dropping from an overhead stalactite. Therefore, whenever a stalagmite is seen, its matching stalactite can usually be identified overhead. Sometimes the separation between the two is only a few inches; in other cases, it can be several hundred feet (the distance between cave floor and cave ceiling). Occasionally, water will drip from the cave ceiling at a rate too rapid for any serious buildup of minerals to occur. The dripping water on the floor, however, may be able to accumulate to a degree whereby a stalagmite can be formed. In such instances, the stalagmite will not have a matching stalactite overhead. Stalagmites encountered in caves that do not have overhead stalactites may be examples of this type of forming pattern, but more often than not the overhead stalactite broke loose from the cave ceiling eons ago and shattered on the floor.

stratum: A bed of rock formed just beneath the earth's surface. The plural is strata, and rock layers are often seen overlapping each other and reflecting different periods in the earth's formation. Water flowing through cracks and fissures in rock strata for many millions of years brought about the formation of the caves we explore today.

straw: A thin, uniform, and hollow stalactite suspended from the cave ceiling. Straw formations are not especially rare, but they are certainly not as common as stalactites and stalagmites. Unlike these formations, straws are not tapered toward their end. They are structurally weak as compared with other types of solid formations, and they rarely attain great lengths without falling from the cave ceiling. Straws a foot long or more are considered a fairly rare find.

subterranean: A term used to describe any physical object or occurrence lying under the earth's surface.

tackle: See *gear.*

travertine: A light-colored rock made of either calcite or aragonite. It is quite common in all caves and is usually very hard. It is composed of microscopic particles deposited for many millions of years from dripping water and natural springs. Travertine is a catch-all term for all types of calcium carbonate deposits. This semicrystalline material will often glow when exposed to carbide lamp illumination.

troglodyte: Any animal that spends its entire life in the total darkness of a cave. It may be used as a slang term to describe spelunkers in general, but more accurately names such creatures as cave fish, blind salamanders, isopods, etc., which spend all their lives underground, never exiting to the surface.

troglophile: A plant or animal that lives most of its life underground, but which does come to the surface occasionally and can exist on the surface, at least for short periods of time. Most bats fall into this category. These creatures can live on the surface, but are better suited and prefer to live underground. Camel crickets are another example; these can live far underground, but will often venture to the surface or to a point near the opening of the cave.

trogloxene: Part-time inhabitants of caves. Most animals and plants in this category are better known as surface-dwellers, but do have some ability to allow them to adapt to a part-time underground environment. Animals falling into this category can include foxes, some birds, amphibians, snakes, etc. Trogloxenes normally use caves for convenience or temporary shelter, and most prefer to live on the surface when the opportunity presents itself.

tunnel: An artificial passageway that is cut through the earth, more or less horizontally, by mining equipment. Some natural passageways which are usually circular and of small diameter may also be referred to as tunnels, although the name is not exactly correct.

twilight zone: In a cave or cavern, that area which lies just inside the entrance and in which some light can de detected. A few feet into the entrance, the light begins to dim. Finally, an area is reached which produces a soft illumination similar to twilight periods on the surface. Naturally, the twilight zone in the cave disappears at night and is usually present only during periods of natural outside light. Some creatures live in the twilight zone area of caves that cannot exist further underground. These are usually surface-dwellers who sometimes take refuge in caves in order to avoid the noontime heat or to hide from predatory animals.

underground lake: A relatively large body of still water. Some lakes are less than 100 feet wide; others may span far greater distances. Underground lakes often reflect the presence of a large depression or well that was filled in by ancient streams. Alternately, it may be a flat area near the botton of a cave that allows incoming streams to collect. Lake is a relative term, being larger than an underground pool. When lakes become tremendously large, they are sometimes known as *underground seas.*

underground pool: A small body of still water which is usually less than 50 feet in diameter. There has been many an argument over the difference between a small underground lake and a large underground pool. The term pool is relative and indicates a still body of water smaller than an underground lake. (see *underground lake*)

underground stream: A flowing body of water which travels through a cave. Most of these streams have origins on the surface, but some may be fed by subterranean springs. Underground streams were responsible for the formation of most caves. They are still carving away at the rock surface in many caves and slowly eking out other passageways.

vanishing stream: A surface stream that suddenly disappears underground. This is caused by the surface water course flowing into an opening or depression which takes the stream underground. Most flowing streams in caves have their origin from a surface stream. Vanishing streams have long been a surface indicator of a hidden cave lying below. Persons seeking out caves look for signs of vanishing streams in conjunction with sinkholes, camel crickets, etc.

vein: A mineral deposit clearly defined in a cave wall, floor, or ceiling. A vein or iron ore, for example, can be clearly seen in limestone deposits because it is a darker color, It closely resembles the veins just beneath the human skin. Some veins are only a fraction of an inch in diameter and run for a length of a foot or so. Others may be several feet wide and travel along a cave wall for great distances.

wash: A deposit of loose sand, gravel, etc., that has been carried by a flowing stream and piled up in the cave. Some deposits are fairly small in size; others may consist of materials deposited over many millions of years. Large washes are responsible for much of the cave fill encountered while spelunking today. (see *cave fill*)

water table: Beneath the surface of the earth, there is a zone which is saturated with water. The water table is the upper surface of this zone and will fluctuate from season to season depending upon the amount of natural rainfall. In the dry summer months, the water table will decrease in content and thus fall to a lower level. It will rise again during the rainy season.

well: A depression in the floor of a cave. It may be called a pit or a chasm, but more often it is a fairly narrow depression that runs more or less vertically to a lower level of the cave. A well is simply a chimney that moves down instead of up.

Index